Study Guide

Christine Stamm-Griffin

ON COOKING
A Textbook of Culinary Fundamentals

Fourth Edition

Sarah R. Labensky • Alan M. Hause
with Steve Labensky and Priscilla Martel

PEARSON
Prentice Hall

Upper Saddle River, New Jersey 07458

Pearson Prentice Hall™ is a trademark of Pearson Education, Inc.
Pearson® is a registered trademark of Pearson plc
Prentice Hall® is a registered trademark of Pearson Education, Inc.

Pearson Education LTD.
Pearson Education Singapore, Pte. Ltd.
Pearson Education Canada, Ltd.
Pearson Education–Japan
Pearson Education Australia PTY, Limited
Pearson Education North Asia Ltd.
Pearson Educación de Mexico, S.A. de C.V.
Pearson Education Malaysia, Pte. Ltd.

10 9 8 7 6 5 4 3 2
ISBN 0-13-171338-8

Dedication

For Linda Cullen, beloved friend, chef and educator whose heart and soul will forever be reflected in the students that she taught. Without Linda's collaboration, this study guide would never have become a reality.

CONTENTS

ABOUT THE AUTHOR

Dr. Christine Stamm-Griffin is an experienced culinarian and Professor at Johnson & Wales University in Denver, Colorado. During her fifteen years teaching at Johnson & Wales she has written and taught a variety of culinary and service courses in the Associate and Baccalaureate degree programs. In addition to serving as a member of the opening team for the Johnson & Wales University Denver campus, she is a primary author of Johnson & Wales' *Culinary Fundamentals* textbook.

Christine has earned an A.O.S degree in Culinary Arts, a B.S. degree in Foodservice Management, and a M.S. degree in Management Technology. Most recently she completed her Ed.D. in Curriculum and Teaching from Boston University's School of Education. In addition, she is certified as an Executive Chef and Culinary Educator by the American Culinary Federation. She has achieved acclaim, winning several national and international gold awards in food show competitions such as Hotelympia in London, England; Chef Ireland; the Culinary Olympics in Frankfurt, Germany; and the New England Chapters of the American Culinary Federation. Outside of teaching she serves as a foodservice consultant with Yum! Inc., of Aurora, Colorado. Christine also conducts professional lectures on topics in culinary arts, culinary nutrition, and education. She is currently the Department Chairperson and a professor for the School of Education at Johnson & Wales, where one of the programs she leads is a B.S. degree in Family & Consumer Sciences Education, a program that prepares candidates to become licensed to teach in high school family and consumer sciences (including foods) classes.

Dr. Stamm-Griffin resides in Colorado with her husband Jim, daughter Haley and son Jonah.

1

PROFESSIONALISM

➢ **T**EST **Y**OUR **K**NOWLEDGE

The practice sets provided below have been designed to test your comprehension of the information found in this chapter. It is recommended that you read this chapter completely before attempting these questions.

1A. Terminology

Fill in the blank spaces with the correct definition.

1. Brigade_____

2. Gourmand _____

3. Nouvelle cuisine_____

4. Sous chef _____

5. Grande cuisine _____

6. Fernand Point_____

7. Classic cuisine _____

8. Executive chef _____

9. Dining room manager _____

10. Marie-Anton Carême _____

11. Line cooks ·_____

12. Auguste Escoffier_____

13. Pastry chef _____

14. Area chef _____

15. Head waiter _____

16. Cuisine Bourgeosie _____

17. New American cuisine_____

18. Fusion cuisine _____

19. Restaurateur _____

1B. Fill in the Blank

Fill in the blanks provided with the response that correctly completes the statement.

1. _____ is terminology used in reference to the foodservice industry to describe the area where guests are generally not allowed, such as the kitchen.

2. _____ service tends to be more formal by using two waiters: a captain and a waiter.

3. _____ is terminology used in reference to the foodservice industry to describe the area where guests are welcome and serviced, such as the dining room.

4. In _____ service, the entree, vegetables and potatoes are served from a platter onto a plate by the waiter.

5. A_____ _____ is responsible for making sure the tables are properly set, foods are delivered in a timely fashion to the proper tables and that the guests' needs are met.

6. In _____ service, one waiter takes the order and brings the food to the table.

7. _____ is another term used for the position of the dining room attendant, who clears plates and refills water glasses at various tables in the dining room.

8. The sommelier, or _____, is responsible for all aspects of wine service.

9. _____ is a tool, delivered to professional kitchens everywhere by the use of modern computer technology, that enables chefs to communicate more effectively and source and order foods from a world of suppliers.

1C. Short Answer
Provide a short response that correctly answers each of the questions below.

1. List three (3) advantages that the introduction of the cast iron stove lent to professional, 19th century cooking.

    a._____

    b. _____

    c._____

2. List three (3) examples of food preservation and storage techniques that were developed in the 19th century.

    a._____

    b. _____

    c._____

3. What is the reasoning behind the design of each element of the professional chef's uniform?

    a. Neckerchief: _____

    b. Black and white checked trousers: _____

    c. White double-breasted jacket _____

    d. Apron: _____

4. List three (3) ways a professional chef can show pride in performing his/her job.

    a._____

    b _____

    c._____

1D. Defining Professionalism

A student chef should try to develop six basic attributes in readiness for his/her role as a professional chef. Fill in the blank with the term that best matches the definition given on the right. (Term choices: knowledge, skill, taste, judgment, dedication and pride)

Term   *Definition*

1. _____   The ability to make sound decisions such as what items to include on the menu; what, how much, and when to order food; and approving finished items for service; all of which can only be learned through experience.

2. _____   The desire to continually strive for the utmost professionalism and quality in spite of the physical and psychological strains of being a chef.

3. _____   A chef's ability to prepare flavorful and attractive foods that appeal to all senses and to the desires of his/her clientele.

4. _____   The desire to show high self-esteem for one's personal and professional accomplishments by means of such details as professional appearance and behavior.

5. _____   An ability developed through practical, hands-on experience that can only be perfected with extended experience.

6. _____   The understanding of a base of information that enables a chef to perform each aspect of the job.

1E. Noteworthy Chefs

Match each of the chefs in List A with the major culinary claim to fame s/he is known for in List B. Each choice in List B can only be used once.

List A   *List B*

_____ 1. Alice Waters   a. Trained by Fernand Point and pioneers of nouvelle cuisine.

_____ 2. Boulanger   b. One of first to offer a menu listing available dishes during fixed hours of operation.

_____ 3. Fernand Point   c. The father of nouvelle cuisine.

_____ 4. Auguste Escoffier   d. Exhibited culinary skill in grande cuisine in some of Europe's finest hotels.

_____ 5. Paul Bocuse, Roger Vergé e. Prepared meals consisting of dozens of courses of elaborately prepared and presented foods.

_____ 6. Antonin Carême   f. First known to offer a variety of foods prepared on premises to customers whose primary interest was dining.

                                     g. Served fresh American food, simply prepared.

1F. Matching

Match each of the terms in List A with the appropriate duty/responsibility in List B.   Each choice in List B can only be used once.

	List A		List B
_____	1. Saucier	a.	Sauteéd items and most sauces
_____	2. Friturier	b.	Chocolate éclairs
_____	3. Potager	c.	Caesar salad
_____	4. Garde Manger	d.	All vegetable and starch dishes
_____	5. Rotisseur	e.	Stocks and soups
_____	6. Poissonier	f.	Grilled veal tenderloin
_____	7. Glacier	g.	Roast pork au jus
_____	8. Grillardin	h.	Ground beef
_____	9. Boucher	i.	French fries
_____	10. Boulanger	j.	Steamed asparagus with hollandaise sauce
_____	11. Entremetier	k.	French bread
_____	12. Commis	l.	Poached sole with caper sauce
		m.	Apprentices

1G. Chapter Review

For each statement below circle either True or False to indicate the correct answer. If an answer is false, then explain why.

1. Aside from quickly preparing foods to order, a short order cook may serve much the same role as a tournant; having mastered many cooking stations.

    True   False

2. Today most well run foodservice operations use the formal kitchen brigade system as the means for organizing the kitchen staff.

    True   False

3. Most new concerns that affect the foodservice industry, such as nutrition and sanitation, are brought about by the government.

    True   False

4. It was not until the early 1900s that advances in transportation efficiency improved to the point where the foodservice industry finally began to expand.

    True   False

5. A regional cuisine is composed of recipes based on local ingredients, traditions and practices.

    True   False

6. The biggest difference between establishments serving buffets is that restaurants charge by the dish whereas cafeterias charge by the meal.

    True   False

7. The process of cooking can be described as transferring energy from the heat source to the food to alter the food's molecular structure.

   True   False

8. Most consumers choose a restaurant or foodservice establishment because it provides quality service and food for a price they are willing to pay.

   True   False

9. Escoffier is credited with developing the kitchen brigade system used in large restaurant kitchens.

   True   False

10. Although sun-drying, salting, smoking, pickling and fermenting are effective means of preserving foods, they were passed up for newer technologies due to the labor intensity of preparation.

   True   False

11. As chefs of classical cuisine, it can be said that Carême, Point and Escoffier also practiced gastronomy.

   True   False

12. Dining in the European or Western style by proceeding through a meal course by course, enables diners to simultaneously satisfy all five major groups of taste: sweet, salty, bitter, sour and spicy.

   True   False

13. Escoffier's most important contribution to culinary arts was *Le Guide Culinaire*, an extensive book of classic garnishes for the professional chef.

   True   False

1H. Putting It All Together

Provide a short response for each of the questions below. These questions are designed to help you connect the bigger concepts presented in this chapter and/or text.

1. What is the value of learning about the evolution of the culinary profession?

2. While Chef Gaston LeNôtre is known for his advancement of the baking and pastry profession, as an owner he was exceptionally entrepreneurial. Compare and contrast in detail what LeNôtre did differently from other famous chefs like Carême and Escoffier, hypothesizing whether or not the different era in which he lived played any role in these differences.

3. This chapter has presented a historical perspective of the development of cookery as a profession. With that in mind, explain what impact the use of modern technologies have had on the production of raw food ingredients. In developing your answer consider touching on the switch from organic to chemical fertilizers and pesticides (and the resurgence of organic produce production); the rise of traditional hybridization techniques as well as genetic engineering; advancements in animal husbandry and aquaculture; commercially raised foods that were once only available in the wild; and modern preservation and transportation methods.

2

FOOD SAFETY AND SANITATION

> ## TEST YOUR KNOWLEDGE

The practice sets provided below have been designed to test your comprehension of the information found in this chapter. It is recommended that you read this chapter completely before attempting these questions.

2A. Terminology

Fill in the blank spaces with the correct definition.

1. Clean _____

2. Contamination _____

3. Intoxication _____

4. Microorganisms _____

5. Pathogenic bacteria _____

6. Atmosphere _____

7. Infection _____

8. Acid/alkali balance _____

9. Direct contamination _____

10. Potentially hazardous foods (PHF) _____

11. Toxin-mediated infection _____

12. Trichinosis _____

13. Temperature danger zone (TDZ) _____

14. Virus _____

15. Parasites _____

16. Anisakiasis _____

2B. Multiple Choice

For each request below, choose the one correct response.

1. Which of the following is *not* a necessity for bacteria to survive and reproduce?
   a. Food
   b. Time
   c. Moisture
   d. Sunlight

2. The federal government enacted legislation designed to reduce hazards in the work area and therefore reduce accidents. This legislation is called:
   a. Safe Jobs for Working Americans Act (SJWAA)
   b. Occupational Hazards Prevention Policy (OHPP)
   c. Safety and Health for Working Americans (SHWA)
   d. Occupational Safety and Health Act (OSHA)

3. Which of the following is *not* an important step in proper hand washing?
   a. Use hot running water to thoroughly wet hands and forearms.
   b. Apply antibacterial soap and rub hands and arms briskly with lather for at least 10 seconds.
   c. Scrub between fingers and under nails with a nail brush.
   d. Rinse thoroughly with hot running water.
   e. Reapply soap and scrub hands and forearms for another 5-10 seconds, then rinse again in hot water.

4. Choose the statement that accurately describes how frozen foods should be defrosted. Pull the product from the freezer and:
   a. Microwave on high in a plastic pan deep enough to catch the moisture.
   b. Thaw at room temperature in a pan deep enough to catch the moisture.
   c. Thaw in a warming oven on a roasting rack.
   d. Thaw under refrigeration in a pan deep enough to catch the moisture.

5. Foods that are considered acidic have a pH that is:
   a. At 7.0
   b. 8.5 to 10.0
   c. 0.0 to below 4.6
   d. 10.0 to 14.0

6. Which *one* of the following factors is most easily controlled by food service workers, therefore limiting bacterial growth?
   a. Food
   b. Time
   c. Moisture
   d. Temperature
   e. pH

2C. Chapter Review

For each statement below circle either True or False to indicate the correct answer. If an answer is false, then explain why.

1. The time-temperature principle is one of the best rules to follow to control the growth of bacteria.
   True   False

2. The first thing that should be done when a pest infestation is discovered is to try to find the source.
   True   False

3. A contaminated food will have an unusual odor.
   True   False

4. When cooling semisolid foods, they may be placed in any size container provided they are refrigerated at 40°F (4°C) or below.
   True   False

5. Food handlers are a major cause for the spread of bacteria.
   True   False

6. A dish can be clean without being sanitary.
   True   False

7. The acronym HACCP stands for Hazard Analysis Critical Control Points.
   True   False

8. Vinyl or plastic gloves are important to food handlers because they eliminate the need to wash hands frequently.
   True   False

9. Hepatitis A is a parasite that often enters shellfish through polluted waters, is carried by humans and is often transmitted either by cross contamination or by infected food handlers practicing poor personal hygiene.

   True   False

10. The high internal temperatures reached during cooking (165° F-212°F/74°C-100°C) kill most of the bacteria than can cause food-borne illnesses.

    True   False

11. Reducing a food's water activity level to below 0.85AW kills all microorganisms.

    True   False

12. One can make an effective sanitizing solution that lasts for up to 2 hours by combining 1 gallon of lukewarm water with 1 tablespoon of chlorine bleach.

    True   False

13. If home cooks follow proper health procedures, most cases of illnesses caused by E.coli and salmonella could be avoided.

    True   False

14. The steps for the two tasting spoon method for sampling foods for finishing touches involves dipping the first spoon into the food product, tasting, discarding the first spoon, then repeating the process with the second spoon.

    True   False

2D. HACCP Overview

1. Number the following HACCP steps (1-8) to indicate the order they should be considered in the flow of a food service operation.

   _____ a. Selecting menu and recipes

   _____ b. Holding and service

   _____ c. Storing

   _____ d. Receiving

   _____ e. Preparing

   _____ f. Cooling leftovers

   _____ g. Cooking

   _____ h. Reheating

2. Which one of the following statements is *false* in regards to HACCP?

   a. It is a rigorous system of self-inspection that ensures food service standards are followed.

   b. Law mandates that all food service operations establish and maintain an HACCP system.

   c. It identifies what actions can be taken to reduce or prevent each risk or hazard.

   d. Hazards must be prioritized and correction of critical concerns should take priority.

3. Which statement is *false* regarding the HACCP system?

    a. It focuses on the flow of food through the food service facility.

    b. It is a rigorous system of sanitary inspection conducted by the Health Department.

    c. It is an effective and efficient method for managing and maintaining sanitary conditions in a food service operation.

    d. It is a system that should be followed on a daily basis.

4. Which two statements are *true* about the HACCP system?

    a. HACCP is best applied only to institutional food service establishments.

    b. Fully check and record time-temperature information in written logs.

    c. All personnel must be constantly aware and responsive to the system, the risks, problems and solutions.

    d. HACCP monitors where a mistake can result in the transmission, growth or survival of viruses, fungi, parasites or putrefactive bacteria.

2E. Food-Borne Diseases Review

This section provides a review of information regarding food-borne diseases. Fill in the blanks provided with the response that correctly completes each portion of the statement. Below is a definition of each of the points needing answers for each question.

Organism: What type of organism causes the disease? Is it a bacteria, parasite, virus, fungi, mold or yeast?

Form: Especially relevant to bacteria, what form does it take? Is it a cell, a toxin or a spore?

Source: In what foods might this organism be found, or what is the source of the contaminant?

Prevention: How can an outbreak of this disease be avoided?

1. Botulism

    Organism:_____

    Form: _____

    Source: _____

    Prevention: _____

2. Hepatitis A

    Organism:_____

    Source: _____

    Prevention: _____

3. Strep

   Organism:_____

   Form: _____

   Source: _____

   Prevention: _____

4. Perfringens or CP

   Organism:_____

   Form: _____

   Source: _____

   Prevention: _____

5. Norwalk Virus

   Organism:_____

   Source: _____

   Prevention: _____

6. Salmonella

   Organism:_____

   Form: _____

   Source: _____

   Prevention: _____

7. E. coli or 0157

   Organism:_____

   Form: _____

   Source: _____

   Prevention: _____

8. Trichinosis

   Organism:_____

   Source: _____

   Prevention: _____

9. Anisakiasis

Organism:_____

Source: _____

Prevention: _____

10. Listeriosis

Organism:_____

Form: _____

Source: _____

Prevention: _____

11. Staphylococcus

Organism:_____

Form: _____

Source: _____

Prevention: _____

2F. Matching

Match each of the food categories in List A with the appropriate internal cooking temperature in List B. Some answers in List B may be used more than once.

List A	*List B*
_____ 1. Pork, ham, bacon	a. 165°F/74°C
_____ 2. Poultry	b. Cook until opaque and firm and shells open
_____ 3. Stuffing, stuffed foods	c. 145°F/63°C for 15 secs. or casseroles
_____ 4. Fish	d. 155°F/68°C for 15 secs.
_____ 5. Ground turkey or chicken	e. 180°F/82°C (thigh), 170°F/77°C (breast)
_____ 6. Commercial game	f. Opaque and flaky or 145°F/63°C for 15 secs.
_____ 7. Egg dishes	g. 145°F/63°C
_____ 8. Beef, veal, lamb steaks;	h. 160°F/71°C chops or roasts
_____ 9. Ground beef, veal, pork	i. 175°F/63°C for 15 secs. or lamb
_____10. Eggs	
_____11. Injected meats	
_____12. Shellfish	

3

NUTRITION

> ## TEST YOUR KNOWLEDGE

The practice sets provided below have been designed to test your comprehension of the information found in this chapter. It is recommended that you read this chapter completely before attempting these questions.

3A. Terminology

Fill in the blank spaces with the correct definition.

1. Essential nutrients _____

2. Ingredient alternatives _____

3. Complex carbohydrates _____

4. Dietary fiber _____

5. Calorie _____

6. Metabolism _____

7. Ingredient substitutes _____

8. Simple carbohydrates _____

9. Nonessential nutrients _____

10. Acid _____

11. Base _____

3B. The Chef's Role in Nutrition

Fill in the blank provided with the response that correctly completes the statement.

1. List five things a food service worker can do to meet the diverse nutritional needs of the consumer.

    a. _____

    b. _____

    c. _____

    d. _____

    e. _____

2. Chefs should be receptive to patron requests that attempt to control calorie and fat intake such as:

    a. _____

    b. _____

    c. _____

    d. _____

3. When modifying a recipe, the chef should first identify the ingredient(s) or cooking method(s) that may need to be changed. Once that is done, what three (3) principles should the chef follow to make the dish healthier?

    a. _____

    b. _____

    c. _____

3C. Roles of Nutrients on Health

Match each of the foods/nutrients in List A with the role it plays in the body in List B. Each choice in List B can only be used once.

List A	*List B*
_____ 1. Lipids	a. An important source of energy for the body
_____ 2. Fiber	b. Help to generate energy from foods we eat
_____ 3. Cholesterol	c. The body produces all it needs
_____ 4. Proteins	d. Necessary for transporting nutrients and wastes
_____ 5. Vitamins & Minerals	e. Provide calories, help carry fat soluble vitamins, and give food a pleasant mouthfeel
_____ 6. Carbohydrates	f. Necessary for manufacturing, maintaining and repairing body tissue, and regulating body processes
_____ 7. Water	g. Keeps digestive track running smoothly
	h. Prevent some forms of cancer and heart disease

3D. Essential Nutrients

For each request below, choose the one correct answer.

1. Saturated fats are usually solid at room temperature and are found in which of the following food sources?
   a. Fruits, vegetables, grains
   b. Canola and olive oils
   c. Corn, cottonseed, sunflower and safflower oils
   d. Milk, eggs, meats and other foods from animal sources

2. One should be physically active everyday, including at least how many minutes of moderate physical activity as a baseline for *maintaining* a healthy weight and lifestyle?
   a. 20
   b. 30
   c. 60
   d. 90

3. Which of the following statements is *false* about complex carbohydrates?
   a. They are naturally occurring in sugars in fruit, vegetables and milk.
   b. Fiber is a complex carbohydrate that cannot be digested.
   c. Complex carbohydrates are digested into glucose.
   d. Starch is a complex carbohydrate.

4. Which of the following is *false* about vitamins? They are:
   a. vital dietary substances needed to regulate metabolism.
   b. can be divided into two categories: fat soluble and water soluble.
   c. necessary for manufacturing, maintaining and repairing body tissue.
   d. non-caloric and needed in small amounts.

5. Which of the following statements is *false*?
   a. 1 gram of pure fat supplies 9 kcals
   b. 1 gram of pure protein supplies 4 kcals
   c. 1 gram of pure vitamins supplies 0 kcal
   d. 1 gram of pure carbohydrates supplies 2 kcals

6. Which one of the following is not considered a food that contains simple carbohydrates?
   a. Fruit
   b. Milk
   c. Oats
   d. Vegetables

7. Which essential nutrient forms a gel type substance in the digestive track and helps to reduce serum cholesterol by removing it from the body?

    a. Soluble fiber
    b. Lipids
    c. Simple carbohydrates
    d. Protein

8. Which statement is *false* about hydrogenation?

    a. It results in the formation of trans fats.
    b. Trans fats contain cholesterol.
    c. Trans fats are considered risk factors for heart disease and possibly other diseases like cancer.
    d. Margarine is a product that has been hydrogenised.

9. Which *one* of the following three organizations is responsible for providing recommendations for planning a diet and lifestyle to enhance health?

    a. USDA
    b. NRA
    c. FDA
    d. ACS

10. It has no calories, tastes 200 to 700 times sweeter than table sugar and at one time was linked to causing cancer in laboratory animals although further testing suggested that it is safe enough for human consumption. This defines which artificial sweetener?

    a. Aspartame
    b. Sunnette or Sweet One
    c. Stevia
    d. Saccharin

3E. Parts of a Food Label

Identify the five areas of importance on the food label below. Briefly explain the significance of each.

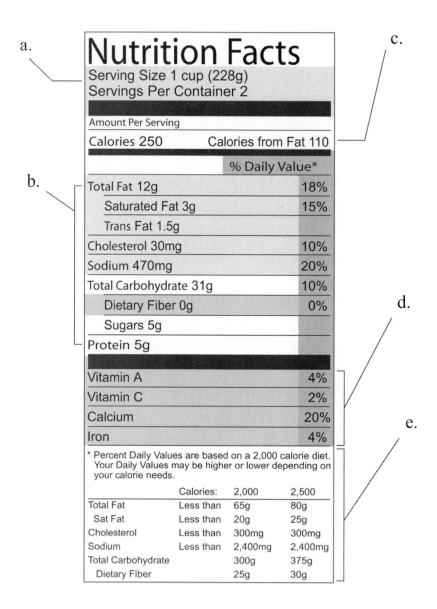

a.

c.

b.

d.

e.

Nutrition Facts

Serving Size 1 cup (228g)
Servings Per Container 2

Amount Per Serving

Calories 250	Calories from Fat 110

	% Daily Value*
Total Fat 12g	18%
Saturated Fat 3g	15%
Trans Fat 1.5g	
Cholesterol 30mg	10%
Sodium 470mg	20%
Total Carbohydrate 31g	10%
Dietary Fiber 0g	0%
Sugars 5g	
Protein 5g	

Vitamin A	4%
Vitamin C	2%
Calcium	20%
Iron	4%

* Percent Daily Values are based on a 2,000 calorie diet.
  Your Daily Values may be higher or lower depending on
  your calorie needs.

	Calories:	2,000	2,500
Total Fat	Less than	65g	80g
Sat Fat	Less than	20g	25g
Cholesterol	Less than	300mg	300mg
Sodium	Less than	2,400mg	2,400mg
Total Carbohydrate		300g	375g
Dietary Fiber		25g	30g

Part of Label   **Significance**

a. _____   _____

b._____   _____

c. _____   _____

d _____   _____

e. _____   _____

3F. The Food Guide Pyramid

Match each of the terms in List A with the appropriate letter in List B. Each choice in List B can only be used once.

List A

_____ 1. Fats, oils, sweets

_____ 2. Proportionality (servings from each group or food band)

_____ 3. Variety

_____ 4. Activity

_____ 5. Food guide pyramid

_____ 6. Personalization

_____ 7. Five basic food group bands

_____ 8. Most recent food pyramid update

List B

a. Symbolizes a personalized approach to healthy eating and physical activity

b. 2005

c. Personal daily physical activity is represented by the steps and the person climbing them

d. Use MyPyramid.gov to personalize the food guide pyramid to you as an individual

e. The thinnest vertical band on the food guide pyramid, suggesting a very small quantity should be consumed each day

f. Represented by the color bands on the pyramid, foods from all groups are needed each day for good health

g. How much food a person should choose from each group or food band

h. Grains, vegetables, fruits, milk, meat and beans

i. 2000

3G. Chapter Review

For each statement below circle either True or False to indicate the correct answer. If an answer is false, then explain why.

1. Restaurateurs are required to supply nutrition information to customers who request it only if they have sold a menu item to the guest about which they've made a nutrition claim such as "low fat."

   True   False

2. Age and gender are the two attributes that determine the different amounts of various nutrients our bodies need.

   True   False

3. The body cannot digest fiber so it passes through the body almost completely unchanged.

   True   False

4. RDA stands for Recommended Daily Allowances.
   True   False

5. Food processing and preparation can reduce a food's mineral content.
   True   False

6. Carbohydrates, proteins and fats can be categorized as energy nutrients.
   True   False

7. A chef's moral and primary responsibility as a foodservice professional is to prepare and serve food that meets or exceeds the guidelines set forth by the Food Guide Pyramid.
   True   False

8. Cholesterol can be found in foods from both plant and animal sources.
   True   False

9. Canola, olive, cottonseed and corn are examples of oils that are a combination of 3 kinds of fat.
   True   False

10. Only 9 of the 20 amino acids are essential.
    True   False

11. Any artificial sweetener can be substituted for sugar when preparing baked goods.
    True   False

12. The six categories of essential nutrients are carbohydrates, lipids, proteins, vitamins, minerals and water.
    True   False

13. Insoluble and soluble fibers are considered simple carbohydrates.
    True   False

14. Proteins can be found in chicken, beef, and salmon as well as beans, rice and pasta.
    True   False

15. A portion of roasted beef tenderloin will likely contain more vitamins than a portion of beef stew due to the different ways they are cooked.
    True   False

16. People are told to lower their cholesterol because they consume too much of it in food and it has a negative effect on bodily functions.
    True   False

17. One thing that all essential nutrients have in common is that they all contribute calories to the body.
    True   False

18. Butter is an example of a hydrogenated fat.
    True   False

3H. Putting It All Together

Provide a short response for each of the questions below. These questions are designed to help you connect the bigger concepts presented in this chapter and/or text.

1. Keep track of everything you eat in one day by recording the information in a notebook. Be sure to include portion sizes with your notes and be completely honest about what you eat. Also note the type of physical activity you get in that day, whether it is low intensity, medium or high and the duration of that exercise.

    Go to MyPyramid.gov and double click on My Pyramid tracker on the left hand side of the web page. Click on the option for "new user" and register on the web page. Once registered, enter all the information you recorded in your notebook regarding your dining patterns and activity for the one day you tracked it.

    Once the analysis is supplied, how would you rate your eating and activity habits? Discuss this with your classmates to find out what variety exists with people who comply/disregard the Food Guide Pyramid recommendations. How could you improve? According to the Food Guide Pyramid what habits do you already possess that are positive?

    Analyze the effectiveness of the Food Guide Pyramid online tool for all Americans.

2. "Eat all foods, but in moderation" illustrates an important concept presented in the Food Guide Pyramid for those who have a love for food. Although some foods may have negative characteristics if consumed regularly or in large quantities, if these foods are eaten in small portions, infrequently, and if the diner gets regular moderate to vigorous daily exercise these foods can actually be considered to be good for us. Provide at least 4 examples of this principle.

4

MENUS AND RECIPES

➢ **TEST YOUR KNOWLEDGE**

The practice sets provided below have been designed to test your comprehension of the information found in this chapter. It is recommended that you read this chapter completely before attempting these questions.

4A. Terminology

Fill in the blank spaces with the correct definition.

1. Entrée _____

2. Static menu _____

3. Cycle menu _____

4. Market menu _____

5. Hybrid menu _____

6. California menu _____

7. Á la carte _____

8. Semi á la carte _____

9. Table d'hôte / Prix fixe _____

10. Recipe _____

11. Standardized recipe _____

12. Weight _____

13. Volume_____

14. Count _____

15. U.S. system _____

16. Metric system_____

17. Yield_____

18. Conversion factor_____

19. As-purchased costs or prices _____

20. Unit costs or prices_____

21. Total recipe cost_____

22. Cost per portion _____

23. Parstock _____

4B. Units of Measure

Fill in the blanks for the following conversions.

1.	1 lb.	=	_____oz.						
2.	1 oz.	=	_____g.						
3.	1 lb.	=	_____g.	=	_____kg.				
4.	1kg.	=	_____g.						
5.	1 g.	=	_____oz.						
6.	1kg.	=	_____oz.	=	_____lb.				
7.	1 c.	=	_____tbsp.	=	_____fl oz.				
8.	2 pt.	=	_____qt.	=	_____fl oz.				
9.	2 qt.	=	_____gal.	=	_____pt.				
10.	2 c.	=	_____tbsp.	=	_____fl oz.				
11.	.5 c.	=	_____tbsp.	=	_____tsp.				
12.	2 c.	=	_____pt.	=	_____qt.				
13.	2 fl oz.	=	_____c.	=	_____tbsp.	=	_____tsp.		
14.	1 gal.	=	_____fl oz.	=	_____qt.	=	_____c.		
15.	1.5 qt.	=	_____pt.	=	_____gal.				

4C. Conversion Factors

Solve the following math problems by calculating the new conversion factors for each example. Note: Round all answers to word problems in sections 4C, 4E, 4F, and 4G to the nearest hundredth decimal point.

1. The original recipe for Tartar Sauce yields 1 gallon, but you need to prepare 2.5 gallons.
    Conversion factor:_____

2. A Chicken Marsala recipe yields 50 portions, however you're only expecting a party of 35.
    Conversion factor:_____

3. You need to make 7 gallons of beef barley soup, but your recipe yields 12 gallons.
    Conversion factor:_____

4. A recipe for Lasagna that yields 8 portions must be increased to yield enough for 20 guests.
    Conversion factor:_____

5. Alone for the evening, you want to make 1 portion of Veal Paprika from a recipe that originally yielded 4.
    Conversion factor:_____

6. The Broccoli Polonaise recipe yields 8 pounds but you need only 6 pounds.

    Conversion factor:_____

7. Twenty-five pounds (8 oz. per serving) of beef stew is the original yield of the recipe, but you only need to produce 15 pounds.

    Conversion factor:_____

    How many portions will 15 pounds of beef stew yield?_____

8. As a part of your mise en place for Sunday brunch you're required to make 5 gallons of pancake batter but the original recipe yields only 2 quarts.

    Conversion factor:_____

4D. Conversion Problems

When making large recipe changes some additional problems may occur. Give a brief description of each problem.

1. Equipment: _____

2. Evaporation: _____

3. Recipe errors: _____

4. Time: _____

4E. Recipe Conversion

The following recipe presently yields 32 6-oz. portions. Calculate the conversion factors and convert the quantities in the recipe to yield 28 6-oz. portions and 84 3-oz. portions.
NOTE: Remember to convert new yields back into pounds, ounces, and quarts or cups.

Cream of Broccoli Soup

	Old Yield 6 qt. (6 lt.) 32 Portions 6 oz. Each	Conversion Factor Yield I	Yield I 28 Portions 6 oz. each	Conversion Factor Yield II	Yield II 84 Portions 3 oz. each
Butter	3 1/2 oz.		_____		_____
Onion	12 oz.		_____		_____
Celery	2 1/2 oz.		_____		_____
Broccoli	3 lb.		_____		_____
Chicken veloute	4 qt.		_____		_____
Chicken stock	2 qt.		_____		_____
Heavy cream	24 oz.		_____		_____
Broccoli florets	8 oz.		_____		_____

4F. Unit Costs

Solve the following math problems by calculating the unit costs for each ingredient.

1. One case of milk costs $38.25. There are nine (9) half gallons in each case. How much does one cup cost?

    Answer: $_____

2. One case of English muffins costs $18.00. There are six (6) packages of twelve (12) muffins in each case. How much does one (1) English muffin cost?

    Answer: $_____

3. One case of olive oil costs $78.00. There are six (6) gallons per case. How much does one quart cost?

    Answer: $_____

4. Five (5) pounds of sliced American cheese cost $12.00. If one slice weighs half an ounce, how much do two (2) slices cost?

    Answer: $_____

5. One case of cooking wine costs $18.08. If four (4) gallons of wine are in each case, what is the cost per ounce of cooking wine?

    Answer: $_____

6. One case of disposable chef hats costs $14.42 and contains twenty-five (25) hats. What is the unit cost of each hat?

    Answer: $_____

7. Twelve rolls of commercial grade paper towels come in one case costing $23.58.   What is the cost per roll of paper towels?

    Answer: $_____

8. One case of heavy cream containing two gallons of product costs $32.16. What is the cost per quart of cream?

    Answer: $_____

4G.   Cost Per Portion

Solve the following math problems by calculating the cost per portion of food product.

1. Sandwich sales for one week are $1,462.50 and 325 sandwiches are sold. How much did each sandwich sell for assuming each sold for the same dollar amount?

    Answer: $ _____

2. The ingredient cost for producing 30 portions of New England boiled dinner includes: potatoes: $15.75; carrots: $12.25; turnips: $10.60; cabbage: $9.85; corned beef: $75.50.

    What is the cost to produce each dinner?

    Answer: $_____

3. One doizen cost $1.00 and the buffet cook uses 3 eggs to make each omlet. What is the cost for the eggs needed in one omelet?

   Answer: $_____

4. In one day a restaurant sells 68 bowls of soup. Sales for soup total $51.00. What is the cost per bowl of soup for each customer?

   Answer: $_____

5. The cost to purchase 25 pounds of ground beef is $56.25. Assuming that 8-oz. hamburgers are made from the full 25 pounds of meat:

   a. What is the cost per pound for the ground meat?   Answer: $_____

   b. What is the cost for the meat used to make each burger?   Answer: $_____

6. A chef purchases a 9.25 pound pork loin for $35.61. As she trims the loin to portion 5 ounce chops, she loses ½ pound of the yield to waste.

   a. What was the cost per pound of meat remaining for portioning?   Answer: $_____

   b. What is the weight of the meat that is available for portioning?   Answer: $_____

   c. How many chops did the trimmed pork loin yield?   Answer: $_____

   d. What was the final cost for each pork chop?   Answer: $_____

7. One case of lettuce contains 28 heads and costs $17.92. Assuming each portion of lettuce for the house salad costs .16 c:

   a. What is the cost per head of lettuce?   Answer: $_____

   b. Approximately how many portions will the case yield?   Answer: $_____

8. In making an ice cream sundae, one gallon of premium vanilla fudge swirl ice cream costs $12.00 and yields 16 portions. Once quart of chocolate sauce costs $3.30 and is portioned in 2 ounce increments. One quart of maraschino cherries costs $10.60 and yields approximately 90 cherries, one per portion.

   What is the total cost of ingredients to produce two sundaes?   Answer: $_____

4H.   Controlling Food Cost

Briefly describe the impact the following areas have on the operational costs of a restaurant:

1. Menu:_____

2. Purchasing: _____

3. Receiving: _____

4. Storing: _____

5. Issuing: _____

6. Standard portions: _____

7. Waste: _____

8. Sales and service: _____

4I. Chapter Review

For each question below circle either True or False to indicate the correct answer. If an answer is false, then explain why.

1. An appetizer, soup or salad followed by an entrée and finished with a dessert, fruit or cheese course is an example of a typical North American meal.
   True   False

2. Portion or balance scales are commonly used in commercial kitchens to accurately determine volumes of ingredients.
   True   False

3. The weight and volume of water, butter, eggs and milk are the same.
   True   False

4. Static, cycle, market or hybrid menus can be purchased á la carte, semi á la carte and/or table d'hôte.
   True   False

5. As a palate cleanser a cheese course may be served after the entrée and before the dessert in the European tradition.
   True   False

6. Volume measurements are generally less accurate than measuring by weight.
   True   False

7. Figuring the portion cost of every item on a plate served to a guest is the primary thing one has to worry about in order to determine the selling price.
   True   False

8. A guest orders a plate of Chicken a la Russe with appropriate accompaniments which add up to total cost of $2.56 to prepare. If the chef is trying to run his kitchen at a 32% food cost, the minimum selling price should be $9.20.
   True   False

4J. Putting It All Together

Provide a short response for each of the questions below. These questions are designed to help you connect the bigger concepts presented in this chapter and/or text.

1. What benefits exist for guests when the chef of a restaurant offers a tasting menu? What benefits exist for the chef/establishment?

2. What impact can straying from following accurate measurements have on the accuracy of costing standardized recipes? Would the results be similar or different if the chef also varied in his/her portion control techniques?

3. What would the results likely be if a chef did not know how to accurately convert recipes to meet varying production needs?

5

TOOLS AND EQUIPMENT

> ## TEST YOUR KNOWLEDGE

The practice sets provided below have been designed to test your comprehension of the information found in this chapter. It is recommended that you read this chapter completely before attempting these questions.

5A. Terminology

Fill in the blank spaces with the correct definition.

1. Carbon steel _____

2. Stainless steel _____

3. High carbon stainless steel _____

4. Ceramic _____

5. Bird's beak knife _____

6. Scimitar _____

7. Whetstone _____

8. Vertical cutter/mixer _____

9. Flat top _____

10. Griddles _____

11. Salamander _____

12. Rotisserie _____

13. Insulated carriers _____

14. Chafing dishes _____

15. Heat lamps _____

16. Work stations _____

17. Work sections _____

5B. Equipment Identification

Identify each of the following items and give a use for each.

Hand Tools

1. Name of item:_____

   Major use:_____

2. Name of item:_____

   Major use:_____

3. Name of item:_____

   Major use:_____

   (insert artwork hs36.ct, offset/grill spatula

4. Name of item:_____

   Major use:_____

   (insert artwork hs44.ct, meat mallet)

5. Name of item:_____

   Major use:_____

Knives

6. Name of item:_____

    Major use:_____

7. Name of item:_____

    Major use:_____

8. Name of item:_____

    Major use:_____

9. Name of item:_____

    Major use:_____

10. Name of item:_____

     Major use:_____

11. Name of item:_____

     Major use:_____

Cookware

12. Name of item:_____

    Major use:_____

13. Name of item:_____

    Major use:_____

14. Name of item:_____

    Major use:_____

15. Name of item:_____

    Major use:_____

16. Name of item:_____

    Major use:_____

17. Name of item:_____

    Major use:_____

    (insert artwork hs75.ct, full hotel pan)--doublecheck ms artwork #--missing #?

18. Name of item:_____

    Major use:_____

19. Name of item:_____

    Major use:_____

20. Name of item:_____

 Major use:_____

21. Name of item:_____

 Major use:_____

22. Name of item:_____

 Major use:_____

Processing Equipment:

23. Name of item:_____

    Major use:_____

24. Name of item:_____

    Major use:_____

Heavy Equipment:

25. Name of item:_____

    Major use:_____

26. Name of item:_____

    Major use:_____

27. Name of item:_____

    Major use:_____

28. Name of item:_____

    Major use:_____

29. Name of item:_____

Major use:_____

5C. Short Answer

Provide a short response that correctly answers each of the requests below.

1. List three (3) of the six (6) requirements for NSF certification of kitchen tools and equipment.

a. _____

b. _____

c. _____

2. Describe four (4) important criteria for evaluation of equipment for kitchen use.

a. _____

b. _____

c._____

d. _____

3. List and describe the three (3) types of metals used in the production of knife blades.

a. _____

b. _____

c._____

5D. Matching

Match each of the pieces of equipment in List A with the appropriate letter definition in List B. Each choice in List B can only be used once.

List A	List B
_____1. Mandolin	a. Can be used in food up to 400°F/204°C
_____2. Refrigerator	b. The metal used most commonly for knife blades
_____3. Candy	c. A metal that holds and distributes heat very well but is also quite heavy
_____4. Salamander	d. Food is placed on a revolving spit
_____5. Copperware	e. An overhead broiler used to brown the top of foods
_____6. Rotisserie	f. A loosely woven cotton fabric used to strain liquids
_____7. Cast iron	g. The metal which is the most effective conductor of heat for cookware
_____8. Cheesecloth	h. Used for food storage, may be walk-in or reach-in
_____9. Tilting skillet	i. A manually operated slicer used for small quantities of fruit and vegetables
_____10. Aluminum	j. A piece of equipment that can be used for frying or braising
	k. A metal that changes color when in contact with acid foods

5E. Fill in the Blank

Fill in the blank with the response that correctly completes the statement.

1. A _____knife is used for general purpose cutting of fruits and vegetables.

2. The part of the knife known as the _____ is the part of the blade that is inside the handle.

3. Short-order and fast-food operations often use a flat metal surface known as a _____on which to cook food.

4. A _____ _____is useful for chopping large quantities of foods to a uniform size.

5. It is advisable to use _____spoons when cooking with nonstick surfaces.

6. A butcher knife is also known as a _____.

5F. Calibrating a Stem-Type Thermometer

Describe the four (4) basic steps required to calibrate a stem-type thermometer:

1. _____

2. _____

3. _____

4. _____

5G. Chapter Review

For each statement below circle either True or False to indicate the correct answer.

1. Stem-type thermometers should be thrown away if they have been dropped.
   True   False

2. Ventilation hoods should be cleaned and inspected by the hotel/restaurant maintenance staff.
   True   False

3. Some hand-made imported pottery may contain lead in the glaze.
   True   False

4. Class B fire extinguishers are the only ones used for fires caused by oil or grease.
   True   False

5. High carbon stainless steel discolors when it comes in contact with acidic foods.
   True   False

6. A steam kettle cooks more slowly than a pot sitting on a stove.
   True   False

7. Most equipment manufacturers voluntarily submit their designs to NSF for certification to show that they are suitable for use in professional food service operations.
   True   False

8. Seamless plastic or rubber parts on food service equipment is important to prevent cracking or splitting over extended use.
   True   False

9. If a fire extinguisher was needed, one rated Class B would be best to eliminate a grease or oil fire.
   True   False

10. Silicone bakeware is only good for baking; putting it in the freezer destroys its flexibility.
    True   False

11. While pans lined with a polymer such as Silverstone or Teflon provide a slippery, nonreactive finish that keeps food from sticking, and thus enables the chef to use less fat in the cooking process, the great deal of care required to keep this coating from chipping, scratching or peeling may not make it an appropriate pan to have in a commercial kitchen.
    True   False

12. Induction burners are portable and favorable because they maintain a safer and cooler cooking environment, however, will not likely become a regular part of the commercial kitchen.
    True   False

13. The size and design of each work station in a kitchen is determined primarily by the menu.
    True   False

5H. Putting It All Together

Provide a short response for each of the questions below. These questions are designed to help you connect the bigger concepts presented in this chapter and/or text.

1. In Chapter One you learned about numerous chefs whose work has served to advance the culinary profession into what it is today. Contrast Alexis Soyer's contribution to those found in Chapter One.

2. Relate your knowledge of food safety and sanitation to determine what you may have to be careful of when shopping for and storing food in an insulated food carrier.

3. Based on your knowledge of food safety and sanitation, describe some steps you can take to avoid chemical contamination of food products in your establishment.

6

KNIFE SKILLS

➢ **TEST YOUR KNOWLEDGE**

The practice sets provided below have been designed to test your comprehension of the information found in this chapter. It is recommended that you read this chapter completely before attempting these questions.

6A. Terminology

Fill in the blank spaces with the correct definition.

1. Whetstone _____

2. Steel _____

3. Chiffonade _____

4. Rondelles/rounds _____

5. Diagonals _____

6. Oblique or roll-cut _____

7. Lozenges _____

8. Butterfly _____

9. Julienne _____

10. Batonnet _____

11. Brunoise _____

12. Small dice _____

13. Medium dice _____

14. Large dice _____

15. Paysanne _____

16. Gaufrette _____

6B. Knife Safety

Briefly describe the six (6) basic steps for knife safety.

1. _____

2. _____

3. _____

4. _____

5. _____

6. _____

6C. Cuts of Vegetables

Draw the following cuts of vegetables to scale and describe their dimensions. On the lines provided below describe any similarities between the strips and the cubes of vegetables.

_____

_____

_____

1. Julienne   4. Brunoise

2. Batonnet   5. Small dice

3. Paysanne   6. Medium dice

6D. Fill in the Blank

Fill in the blank with the response that correctly completes the statement.

1. There are _____ methods of cutting, one where the _____ acts as the fulcrum and the other where the _____ acts as the fulcrum.

2. Parsley and garlic should be chopped with one hand flat on the _____ of the knife, using a _____ motion.

3. When cutting food always cut _____ from yourself and never cut on _____, _____ or _____ surfaces.

4. When using a whetstone, start by placing the _____ of the knife on the stone. Start sharpening on the _____ side of the stone and finish with the _____ side.

5. An onion is diced by cutting it in half and then making incisions toward the _____ of the onion, without cutting through it.

6E. Dicing an Onion

Describe the five (5) steps necessary to dice an onion.

1. _____

2. _____

3. _____

4. _____

5. _____

6F. Chapter Review

For each statement below circle either True or False to indicate the correct answer. If an answer is false, then explain why.

1. A sharp knife is more dangerous than a dull one.

   True   False

2. Tourner means "to turn" in French.

   True   False

3. A steel is used to sharpen knives.

   True   False

4. Batonnets are also referred to as allumettes.

   True   False

5. A whetstone should be moistened with a mixture of water and mineral oil.

   True   False

6. Paysanne can be a half-inch dice that has been cut in half.

   True   False

7. One should not attempt to catch a falling knife.

   True   False

8. Knives should not be washed in the dishwasher.

   True   False

6G. Putting It All Together

Provide a short response for each of the questions below. These questions are designed to help you connect the bigger concepts presented in this chapter and/or text.

1. In Chapter One you learned about professionalism. How would applying excellent knife skills to menu preparation help to differentiate the final foods/dishes of a professional chef from a practitioner?

2. In Chapter Two you learned about food safety and sanitation. What are some practices you can employ to prevent cross contamination of your knives?

7

FLAVORS AND FLAVORINGS

➤ TEST YOUR KNOWLEDGE

The practice sets provided below have been designed to test your comprehension of the information found in this chapter. It is recommended that you read the chapter completely before attempting these questions.

7A. Terminology

Fill in the blank spaces with the correct definition.

1. Herbs _____

2. Table salt _____

3. Shortenings _____

4. Spices _____

5. Vinegar _____

6. Smoke point _____

7. Pickles _____

8. Flavorings _____

9. Rancid _____

10. Relish _____

11. Taste _____

12. Aroma _____

13. Mouthfeel _____

14. Palate _____

15. Seasoning _____

16. Condiment _____

17. Flavor _____

18. Taste _____

19. Flash point _____

20. Brandy _____

21. Wine _____

22. Liqueur _____

23. Beer _____

24. Liquor _____

25. Vintner _____

26. Viniculture _____

27. Viticulture _____

28. Fermentation _____

7B. Discovering Tastes

Match each of the clues in List A with the appropriate definition or explanation in List B. Each choice in List B can only be used once.

List A

_____ 1. Western definition of taste

_____ 2. Sweet

_____ 3. Umami

_____ 4. The Chinese 5-taste scheme

_____ 5. Bitter

_____ 6. Salty

_____ 7. Sour

List B

a. Maintaining the proper balance of tastes in a dish or during an entire meal assists in maintaining good health and fortune

b. Found in acidic foods, it can vary in intensity but can be made more palatable by adding varying amounts of sweet

c. Helps to finish a dish, heightening or enhancing other flavors, it may occur naturally in the food or be added by the cook

d. Based more on science, it identifies four tastes: sweet, sour, salty, bitter and sometimes umami

e. The practice of arranging tastes on a continuum, rating them as primary or secondary, including sweet, salty, bitter, pungent, harsh and astringent

f. Less preferred across cultures than other tastes, it is potent and easily unbalanced by other tastes like sour or salty

g. Literally means delicious, it is naturally occurring in foods that contain amino acid glutamates such as soy sauce, cheese, meats, mushrooms and tomatoes to name a few.

_____8. Ayurvedic medicine

h. Created by naturally occurring sugars that can be enhanced by small amounts of sour, bitter or salty tastes

i. The Indian way of creating dishes with the balance of 6 tastes that are based on the tastes of various herbs and spices

7C. Categorizing Flavorings

Based on the two categories given, identify the items from the list below that are examples of each category. Fill in the blanks provided under each category heading with the corresponding examples.

Herb

1. _____
2. _____
3. _____
4. _____
5. _____

Spice

6. _____
7. _____
8. _____
9. _____
10. _____
11. _____

paprika

cilantro

capers

black pepper

oregano

thyme

ground mustard

garlic

lemon grass

lavender

coriander

7D. Herbs and Spices

For each statement below, choose the one correct response.

1. Which of the following is not one of the three guidelines to follow when experimenting with the use of different herbs and spices in various dishes?

    a. Flavorings should be added at the beginning of the preparation.

    b. Flavorings should not hide the taste or aroma of the primary ingredients.

    c. Flavorings should be combined in balance, so as not to overwhelm the palate.

    d. Flavorings should not be used to disguise poor quality or poorly prepared products.

2. Which spice does the following description identify?
    Thin layers of bark that are peeled from branches of small evergreen trees and dried in the sun. This pale brown spice is most commonly purchased ground since it is difficult to grind.

    a. Nutmeg

    b. Allspice

    c. Cinnamon

    d. Mace

3. Which spice does the following description identify?
   Hand-picked, dried stigmas of a type of crocus that are the most expensive spice in the world.

   a. Turmeric
   b. Saffron
   c. Poppy seeds
   d. Juniper

4. Which herb does the following description identify?
   Hollow, thin, grass-like stems that have a mild onion flavor and bright green color.

   a. Chervil
   b. Lemon grass
   c. Dill
   d. Chives

5. Which herb does the following description identify?
   A flowering herb commonly used as a flavoring in Mediterranean cooking and having a flavor similar to thyme, only sweeter. The wild version of this herb is known as oregano.

   a. Cilantro
   b. Lemon thyme
   c. Rosemary
   d. Marjoram

6. Which spice does the following description identify?
   Round and beige seeds from the cilantro plant that have a sweet, spicy flavor and strong aroma.

   a. Coriander
   b. Cardamom
   c. Fenugreek
   d. Cumin

7. Which spice does the following description identify?
   A root that comes from a tall, flowering tropical plant and has a fiery yet sweet flavor, with hints of lemon and rosemary. It is used extensively in Asian cookery.

   a. Turmeric
   b. Cloves
   c. Ginger
   d. Caraway

8. Which herb does the following description identify?
   Commonly used in Mediterranean cuisines, it has a strong, warm and slightly peppery flavor with a hint of cloves. It is available in a variety of "flavors"--cinnamon, garlic, lemon, and chocolate.

   a. Garlic chives
   b. Sweet basil
   c. Rosemary
   d. Oregano

9. Which herb does the following description identify?
   Typically used in poultry dishes, with fatty meats or brewed as a beverage, its strong balsamic/camphor flavor does not blend well with other herbs.
   a. Savory
   b. Tarragon
   c. Thyme
   d. Sage

10. Which spice does the following description identify?
    Perhaps the world's oldest spice that is a small, crescent-shaped brown seed with the peppery flavor of rye.
    a. Caraway
    b. Cardamom
    c. Coriander
    d. Mustard

11. Which herb does the following description identify?
    A member of the parsley family that has delicate blue-green, feathery leaves and whose flavor is similar to parsley, only with a hint of anise.
    a. Fennel
    b. Dill
    c. Tarragon
    d. Chervil

12. Which spice does the following description identify?
    An American combination of spices—oregano, cumin, garlic and other flavorings—intended for use in Mexican dishes.
    a. Paprika
    b. Chile powder
    c. Commercial chili powder
    d. Grains of paradise

13. Which spice does the following description identify?
    The dried green leaf of the sassafras plant used by Choctaw Indians as a thickener and flavoring in Cajun and Creole cuisines.
    a. Garlic
    b. Paprika
    c. Horseradish
    d. Filé powder

14. Which spice does the following description identify?
    A pale green root with a strong aroma and sharp cleansing flavor with herbal overtones that is similar to, but hotter than, the unrelated horseradish root.
    a. Turmeric
    b. Galangal
    c. Wasabi
    d. Ginger

15. Which herb does the following description identify?
    Tough, glossy leaves with a sweet balsamic aroma and peppery flavor also known as sweet laurel.

    a. Epazote

    b. Bay

    c. Sage

    d. Lavender

7E. Short Answer

Provide a short response that correctly answers each of the requests below.

1. Even though umami is a relatively new addition to the list of basic tastes for Westerners, it has been part of what country's cuisine focus and taste profile for years?

2. Name three taste receptors in the mouth that help us perceive taste:

    a._____

    b._____

    c._____

3. Without _____the taste compounds cannot be dissolved, enabling them to then stimulate the taste receptors.

4. There are two ways that we smell foods, which enhance our ability to taste, and both are enabled by olfactory bulbs located in two different places. In simple terms, not the complex ones provided in vocabulary in the text, explain where those sense organs are that enable us to smell.

    a._____

    b. _____

5. There are factors affecting one's perception of flavors that a chef must take into account during preparation. Next to the factor listed on the left, write the rule of what the chef should watch out for.

    a. Temperature _____

    b. Consistency:_____

    c. Presence of contrasting tastes:_____

    d. Presence of fats: _____

    e. Color: _____

6. Considering some of the things that can compromise one's perception of taste, why might it be more challenging to be a chef in a nursing home compared to a chef in a restaurant?

    _____

    _____

    _____

7. Describe the process of making sparkling wines, including both the first and second stages of production.

_____

_____

_____

_____

8. Explain what caused an unusually large number of French vintners to relocate throughout Europe, Australia and North America in the late 1800s.

    _____

    _____

    _____

    _____

9. List and explain the four guidelines a chef should use when matching beer and food.

    a._____

    b. _____

    c._____

    d. _____

10. Beer, Brandies, Liquors and Liqueurs

    Fill in the missing information in the following table:

Alcoholic Beverage	Production Method	Base Ingredient	Additional Ingredients
a. German Beer			
b. Cognac			
c. Gin			
d. Rum			
e. Vodka			
f. Tequila			
g. Whiskey			
h. Liqueurs			

7F. Chapter Review

For each statement below, circle either True or False to indicate the correct answer. If an answer is false, then explain why.

1. Ketchup originally referred to any salty extract from fish, fruits or vegetables.

   True   False

2. In terms of making beverages, the term steeping means mixing hot water with ground coffee.

   True   False

3. When preparing a recipe that calls for fresh herbs, the rule to follow when fresh herbs are unavailable is: use more dried herbs than the original fresh variety.

   True   False

4. Mustard never really spoils, its flavor just fades away.

   True   False

5. Vegetable oils are cholesterol free, are virtually odorless and have a neutral flavor.

   True   False

6. Olive oil is extracted from a fruit.

   True   False

7. Distilled vinegar is made from white wine and is completely clear with a stronger vinegar flavor and higher acid content than most vinegars.

   True   False

8. Salt is used as a basic seasoning universally and its flavor can be tasted and smelled easily.

   True   False

9. Every culture tends to combine a small number of flavoring ingredients so frequently and so consistently that they become a definite part of that particular cuisine.

   True   False

10. Once ground, spices lose their flavors rapidly so it is better to purchase them in their whole form, then grind them as needed.

    True   False

11. For thousands of years the primary purpose of spices has been to season foods.

    True   False

12. One can experience certain taste qualities on only certain areas of the tongue. For example, sweetness can only be experienced on the tip.

    True   False

13. Pungent, hot, spicy, piquant or astringent are not technically tastes because they are not detected solely by the taste buds.

    True   False

14. It is possible to scientifically measure one's ability to taste.

    True   False

15. Savory herbs such as dill and basil have no role in the bakeshop.

    True   False

16. All Champagnes are sparkling wines but not all sparkling wines are Champagnes.

    True   False

17. Humans have been consuming wines since the early 1800s and the discovery of the wine making process was quite by accident—air-borne yeasts came into contact with stored grapes or grape juice and over time the mixture fermented, producing a sweet alcoholic beverage.

    True   False

18. Some of the grapes generally recognized as noble used for red wines include Cabernet Sauvignon, Merlot, Chianti, Nebbiolo and Pinot Noir.

    True   False

19. Some of the grapes used for white wines that are also noble include Chardonnay, Riesling and Sauvignon Blanc.

    True   False

20. Alcohol contributes less to the flavor of a wine than the Vintner's balance between the sugars and acids.

    True   False

21. Each grape varietal used to make a wine possesses a certain hallmark aroma and flavor and therefore a Pinot Noir wine produced in California will be almost identical to one produced in Australia.

    True   False

22. Wines have an alcoholic content of 10 to 15% whereas fortified wines contain 18 to 22% by volume.

    True   False

23. The only rule about matching wine with food is that there are no absolute rules.

    True   False

24. The best type of wines to use for cooking are cooking wines.

    True   False

25. Unlike wine, beer does not improve with age and is best consumed as soon as possible after production.

    True   False

26. Of the two main categories of beers produced in the world, lagers are characteristically light, clear and crisp and consumed mainly in Britain and Belgium while ales are aromatic, and cloudy.

    True   False

7G. Putting It All Together

Provide a short response for each of the questions below. These questions are designed to help you connect the bigger concepts presented in this chapter and/or text.

1. A repeat customer walks into a chef's restaurant. While greeting her, the chef learns that the customer has been told by her doctor to cut back on the amount of meat she consumes each day due to her high cholesterol. The customer is frantic, as she loves the taste of meat and she cannot imagine missing that taste in her foods. What ingredients could the chef substitute that would still provide the meaty or delicious taste of meat yet with less cholesterol?

2. Combining your newfound knowledge of taste with your knowledge about nutrition from Chapter Three, explain why a bowl of ice milk may not be as satisfying as a bowl of ice cream.

8

DAIRY PRODUCTS

> ## TEST YOUR KNOWLEDGE

The practice sets provided below have been designed to test your comprehension of the information found in this chapter. It is recommended that you read this chapter completely before attempting these questions.

8A. Terminology

Fill in the blank spaces with the correct definition.

1. Cream cheese _____

2. Heavy whipping cream _____

3. Fondue _____

4. Dairy products _____

5. Non-fat milk_____

6. Half-and-half cream _____

7. Whipped butter_____

8. Skim milk_____

9. Buttermilk _____

10. Salted butter _____

11. Lowfat milk _____

12. Yogurt _____

13. Light cream _____

14. Sour cream _____

8B. Comparing Creams

Match each type of cream in List A with the appropriate fat content in List B. Each choice in List B can only be used once.

List A

_____ 1. Light whipping

_____ 2. Light cream

_____ 3. Half-and-half

_____ 4. Heavy (whipping) cream

List B

a. Not less than 36% milk fat

b. 10%-18% milk fat

c. 16%-23% milk fat

d. 18%-less than 30% milk fat

e. 30%-36% milk fat

8C. Cheese Identification

Match each of the cheese varieties in List A with the appropriate letter definition in List B. Each choice in List B can only be used once.

List A

_____ 1. Mozzarella

_____ 2. American cheddar

_____ 3. Parmigiano-Reggiano

_____ 4. Gruyère

_____ 5. Boursin

_____ 6. Roquefort

_____ 7. Chevre

_____ 8. Brie

_____ 9. Feta

_____ 10. Havarti

_____ 11. Ricotta

_____ 12. Monterey jack

List B

a. French, semi-soft, blue-veined sheep's milk cheese containing 45 % fat.

b. A hard, cow's milk cheese containing from 32% to 35% fat and produced exclusively near Parma, Italy.

c. A cheddar-like cow's milk cheese from California containing 50% fat.

d. A fresh, soft, Italian cow's milk cheese similar to cottage cheese containing 4%-10% fat.

e. A fresh, firm, Italian cow's milk cheese very mild in flavor, that can become elastic when cooked.

f. A French, rindless, soft, triple cream cow's milk cheese usually flavored with garlic, herbs or peppers.

g. A firm, cow's milk cheese made primarily in NY, WI, VT, and OR, 45% to 50% fat.

h. A well-known, mild Wisconsin cheddar containing 45%-50% fat.

i. A sharp-flavored, hard sheep's milk cheese from Central and Southern Italy containing 35% fat.

j. A milder, French or Belgian, soft rind-ripened cheese made from cow's milk, containing 45% fat.

k. A semi-soft cow's milk cheese from Piedmont, Italy containing 45% fat.

l. A soft, French, rind-ripened cheese made with cow's milk and containing 60% fat.

_____13. Gorgonzola

_____14. Mascarpone

_____15. Colby

_____16. Camembert

_____17. Pecorino-Romano

m. A firm, Swiss, cow's milk cheese that is highly flavorful, sweet and nutty, and aged up to 12 months.

n. A fresh, soft, Italian cow's milk cheese originally from Lombardy, Italy containing 70%-75% fat.

o. A fresh, soft, American cow's milk cheese containing 35% fat.

p. A semi-soft, Italian, blue-veined cow's milk cheese containing 48% fat.

q. A pale yellow, Danish cow's milk cheese with many small, irregular holes, often made with herbs and spices.

r. A fresh, Italian or Greek, sheep and/or goat's milk cheese that is white and flaky, from pickling in brine.

s. A soft, creamy, goat's milk cheese with a short shelf life.

8D. Milk Products

For each request below, choose the one correct response.

1. Milk products should be kept refrigerated at or below what temperature?
   a. 30ºF
   b. 35ºF
   c. 40ºF
   d. 45ºF

2. Aside from increasing the shelf life of cream, the process of ultrapasteurization:
   a. reduces the whipping properties
   b. thickens the consistency
   c. causes the cream to stay whipped for longer periods of time
   d. concentrates the fat content

3. Which one of the following is *false* regarding concentrated or condensed milk products? They:
   a. do not require refrigeration once opened.
   b. are produced by using a vacuum to remove all or part of the water from whole milk.
   c. have a high concentration of milkfat and milk solids
   d. have an extended shelf life.

4. Grades of milk are assigned based on:
   a. the clarity of color and distribution of fat globules.
   b. bacterial count; no less than 20 and no more than 30 per gallon earns a Grade A.
   c. bacterial count; the lower it is, the higher the grade.
   d. the flavor of the milk as determined by the breed and feed of the animal.

5. Which one of the following is *false* regarding homogenization? Homogenization:

   a. breaks the fat globules in the whole milk into a smaller size and permanently disperses them.

   b. is not required, but is commonly performed on commercial products.

   c. results in a milk product with a whiter color and richer taste.

   d. increases the shelf life of the milk product.

6. Which of the following is *true* regarding pasteurization? It requires holding the milk at a temperature of:

   a. 140°F for 15 seconds

   b. 161°F for 15 seconds

   c. 275° F for a very short time

   d. 280° F to 300°F for 2 to 6 seconds

7. Evaporated milk, sweetened condensed milk and dry milk powders are all examples of:

   a. canned milk products

   b. concentrated milk products

   c. cultured dairy products

   d. substandard milk products

8. Which of the following is **true** about sweetened condensed milk? It:

   a. contains between 60% and 65% sugar.

   b. can be substituted for whole milk or evaporated milk.

   c. is concentrated like evaporated milk by removing 60% of the water.

   d. has a brilliant, white color and faint flavor of caramel.

9. Dairy products are naturally high in all of the following *except*:

   a. carbohydrates

   b. vitamins

   c. proteins

   d. minerals

10. An enzyme is often used, such as rennet, to coagulate milk proteins in cheese production, therefore separating it into what *two* products?

    a. Liquid fats

    b. Solid curds

    c. Liquid whey

    d. Water

11. In order to produce 3 pounds of the Ricotta cheese using the recipe found in this chapter, the conversion factor would be:

    a. 5

    b. .16

    c. 7.5

    d. 6

12. According to the recipe found in this chapter, 4 quarts of milk and 12 fluid ounces of fresh lime juice will yield how much Ricotta cheese?

    a. 30 ounces
    b. 1.5 pounds
    c. 2 pounds
    d. 40 ounces

13. Which one of the following choices *does not* equal 4 quarts?

    a. 1 gallon
    b. 8 pints
    c. 136 ounces
    d. 16 cups

14. Twelve fluid ounces are equal to:

    a. 1 pint
    b. 1 cup
    c. .5 cup
    d. 1.5 cups

15. Fresh limes cost $2.50 per pound and on the average, eight limes weigh one pound. While squeezing the fresh lime juice for the Ricotta cheese recipe, the chef uses sixteen limes. What is the cost of the lime juice?

    a. $2.50
    b. $4.50
    c. $5.00
    d. $4.25

8E. Chapter Review

For each statement below circle either True or False to indicate the correct answer. If an answer is false, then explain why.

1. Milk products processed by ultra high temperature (UHT) processing can be stored without refrigeration for at least three months.

    True   False

2. Margarine contains cholesterol.

    True   False

3. Coffee whiteners, imitation sour cream, and whipped topping mixes are made from nondairy products.

    True   False

4. All grades of milk must be pasteurized before retail sale.

    True   False

5. The lack of moisture in dry milk powder prevents the growth of microorganisms.

    True   False

6. Seasoning butter with salt changes the butter's flavor and extends its shelf life.

   True   False

7. Both butter and margarine contain about 80% fat and 16 % water.

   True   False

8. Yogurt is a good example of a health or diet food.

   True   False

9. Margarine is a dairy product that serves as a good substitute for butter.

   True   False

10. Aside from excess moisture, processed cheese foods are as equal in quality as natural cheeses.

    True   False

11. One pound of whole butter that is clarified will result in 12 ounces of clarified butter.

    True   False

12. Natural cheeses contain cholesterol.

    True   False

13. Imitation and artificial dairy products may be useful for persons with allergies or on a restricted diet.

    True   False

14. Fluid milk is a potentially hazardous food and should be kept refrigerated at 45°F /5°C.

    True   False

15. Freezing cultured milk products is generally not recommended.

    True   False

16. Dry milk powder and sweetened condensed milk are both examples of concentrated milks.

    True   False

17. The Food and Drug Administration does not allow the manufacture and distribution of rawmilk cheeses in the United States.

    True   False

8F.   Putting It All Together

Provide a short response for each of the questions below. These questions are designed to help you connect the bigger concepts presented in this chapter and/or text.

1. Considering what you learned in Chapter Three, which would add more creamy texture to a Poulet Sauce if used as a final liaison, traditional (American style) whole butter or European-style whole butter? Why?

2. Why do low-fat milks and lower fat creams contain smaller quantities of cholesterol than their full fat counterparts? Refer to Chapter Three for clues to answer this question.

9

MISE EN PLACE

➤ TEST YOUR KNOWLEDGE

The practice sets provided below have been designed to test your comprehension of the information found in this chapter. It is recommended that you read this chapter completely before attempting these questions.

9A. Terminology

Fill in the blank spaces with the correct definition.

1. Steeping _____

2. Clarification _____

3. Sachet_____

4. Meal _____

5. Fresh bread crumbs _____

6. Standard breading procedure _____

7. Onion piquet _____

8. Blanching_____

9. Dry bread crumbs_____

10. Rub_____

11. Paste _____

12. Bouquet Garni _____

13. Shocked/refreshed _____

9B. Short Answer

Provide a short response that correctly answers each of the questions below.

1. How can one keep his/her hands from becoming coated with breading during the standard breading procedure?

   _____ _____

2. What is the procedure for battering foods?

   a._____

   b. _____

   c._____

3. If one pound of butter costs $2.69 and approximately 80% of the whole butter is fat, 16% is water and 4% is milk solids, what cost proportion (of the $2.69) is comprised by the :

   a. Fat: _____

   b. Water:_____

   c. Milk solids: _____

4. When preparing mise en place, a chef should consider what ingredients need to be prepped and should also include:

   a._____

   b. _____

5. List, in proper sequence, the four (4) steps for the standard breading procedure.

   a._____

   b. _____

   c._____

   d. _____

6. If one pound of butter is made up of 80% fat, 16% water and 4% milk solids, how many ounces of the pound of whole butter does each of those components represent?

   a. Fat: _____

   b. Water:_____

   c. Milk solids: _____

9C. Multiple Choice

For each request below, choose the one correct response.

1. If one pound of whole butter costs $2.69, what is the final cost to produce one pound of clarified butter?

   a. $2.69

   b. $3.36

   c. $2.02

   d. $0.67

2. Toasting of nuts and spices does everything except:
   a. Remove rancidity
   b. Brown
   c. Make the flavor more pronounced
   d. Make them crispier and crunchier

3. After processing dried bread crumbs, they should be passed through a tamis to:
   a. Remove the crusts
   b. Enhance the flavor
   c. Remove foreign particles
   d. Ensure even size

4. Which of the following is not an important consideration when determining mise en place needs?
   a. The person completing the final task
   b. Thinking about each task and the most efficient way to complete it
   c. Knowing how long before completion of task should begin
   d. Planning to eliminate unnecessary steps and conserving resources

5. Rehydrating dried fruits or mushrooms is performed using which technique?
   a. Clarifying
   b. Steaming
   c. Toasting
   d. Steeping

6. Modern marinades impart flavor to foods but are also known for their ability to:
   a. Clarify
   b. Color
   c. Tenderize
   d. Preserve

7. A liquid with a medium viscosity made by combining milk, flour, seasonings and baking powder would be a description of:
   a. Slurry
   b. Roux
   c. Batter
   d. Béchamel sauce

8. When considering whether or not to purchase a certain convenience product, the chef must determine all of the following *except*:
   a. Whether he/she can properly store the product
   b. The amount of employee time saved producing the product in house
   c. The cost of the convenience product
   d. The quality and consistency of the convenience product

9. All of the following are true statements regarding shocking or refreshing *except*:
   a. Quickly cools hot foods to a temperature below 41°F/4°C
   b. Will occur more quickly in a plastic container
   c. Should be conducted in a combination of water and ice
   d. Helps maintain the delicate textures and doneness of blanched or par-cooked foods

10. Below are several statements regarding marinating, which one is *false*?
   a. The type of wine used in a wine-based marinade is not an important consideration for flavor.
   b. Cover the food completely with the marinade and refrigerate.
   c. Smaller pieces of food take less time to marinate than larger pieces.
   d. If there is not enough marinade to completely cover the food, heavy-duty plastic food storage bags may be used to keep the marinade in contact with the food product more completely.

9D. Chapter Review

For each statement below circle either True or False to indicate the correct answer. If an answer is false, then explain why.

1. A standard sachet consists of peppercorns, bay leaves, parsley stems, thyme, cloves and optionally, garlic.

   True   False

2. The only purpose for beer in a batter is flavor.

   True   False

3. Mise en place is a task only important for the back of house personnel to complete.

   True   False

4. A food's weight is equal to its volume.

   True   False

5. The composition of Ghee is identical to clarified butter.

   True   False

6. Convenient location of sanitizing solution, hand towels, disposable gloves and trash receptacles should be part of the mise en place planning process.

   True   False

7. Dried bread crumbs should be stored in a tightly closed plastic container in a cool, dry place.

   True   False

8. Breaded foods are usually cooked by roasting or stewing.

   True   False

9. How foods are prepped for mise en place is as important as the way they are stored (at proper temperatures) before the final preparation occurs.

   True   False

10. You've just finished making fresh bread crumbs and they're sticking together in clumps rather than being a smooth mixture of evenly chopped crumbs. Chances are the problem is your bread was stale.

    True   False

11. Partially cooking vegetables as part of mise en place may be done to remove bitter flavors, loosen peels, soften firm foods and shorten final cooking times.

    True   False

12. The combination of water and ice will chill foods more rapidly than a container of only tap water or only ice.

    True   False

9E. Putting It All Together

Provide a short response for each of the questions below. These questions are designed to help you connect the bigger concepts presented in this chapter and/or text.

1. Referring back to Chapter Five on tools and equipment, why are dried breadcrumbs passed through a tamis after processing? In your answer explain what a tamis is and the product that will result by using this piece of equipment.

2. When marinating raw meats in the refrigerator, aside from storing them in a covered container, where should they be placed on the shelving unit? Refer back to Chapter Two, Safety and Sanitation if assistance is needed in answering this question.

10

PRINCIPLES OF COOKING

> ## TEST YOUR KNOWLEDGE

The practice sets provided below have been designed to test your comprehension of the information found in this chapter. It is recommended that you read the chapter completely before attempting these questions.

10A. Terminology

Fill in the blank spaces with the correct definition.

1. Convection _____

2. Coagulation _____

3. Gelatinization_____

4. Radiation _____

5. Conduction_____

6. Combination cooking methods _____

7. Mechanical convection _____

8. Infrared cooking_____

9. Carmelization_____

10. Moist heat cooking methods _____

11. Microwave cooking _____

12. Evaporates _____

13. Melt _____

14. Dry heat cooking methods _____

15. Natural convection_____

10B. Cooking Methods

Fill in the blanks provided with response that correctly completes information about each cooking method.

Cooking method	Medium	Equipment
ex: Sautéing	Fat	Stove
1. Stewing	_____	_____
2. Deep-frying	_____	_____
3. Broiling	_____	_____
4. Poaching	_____	_____
5. Grilling	_____	_____
6. Simmering	_____	_____
7. Baking	_____	_____
8. Roasting	_____	_____
9. Steaming	_____	_____
10. Braising	_____	_____

10C. Smoking Points

Match each of the lipids in List A with the smoke point in List B. Each choice in List B can only be used once.

List A	List B
_____1. Whole butter	a. 410°F/210°C
_____2. Deep-fryer shortening	b. 450°F/232°C
_____3. Corn oil	c. 335°-380°F/168°-193°C
_____4. Lard	d. 410°-430°F/210°-221°C
_____5. Peanut oil	e. 495°F/257°C
_____6. Extra virgin olive oil	f. 260°F/127°C
_____7. Clarified butter	g. 250°F/121°C
_____8. Margarine	h. 370°F/188°C
	i. 440°F/227°C

10D. Multiple Choice

For each question below, choose the one correct response.

1. Which method refers to the transfer of heat through a fluid?

    a. Natural convection

    b. Radiation

    c. Conduction

    d. Induction

2. What cooking technique is an example of moist cooking?

    a. Grilling

    b. Sautéing

    c. Deep-fat frying

    d. Steaming

3. The purpose for the shape of the wok used for stir-frying is that the rounded shape:

    a. Makes it easier to pour liquids out of it.

    b. Is designed to fit into the specially designed shape of the turbo gas burners.

    c. Diffuses the heat and makes tossing and stirring easier.

    d. Makes the cookware more durable.

4. Which of the following is an example of infrared cooking?

    a. Broiling

    b. Sautéing

    c. Roasting

    d. Baking

5. In pan-frying, how much fat or oil should be in the pan?

    a. Just enough to coat the bottom of the pan

    b. 1 cup measure

    c. 1/2 to 2/3 way up on the product being cooked

    d. Enough to completely cover the product

6. What cooking technique is defined in the following statement: "To briefly and partially cook a food in boiling water or hot liquid."

    a. Boiling

    b. Blanching

    c. Frying

    d. Simmering

7. Which does *not* describe convection heat transfer?

    a. The natural tendency of warm liquids and gases to rise while cooler ones fall

    b. Fans or a stirring motion circulate heat

    c. A combination of conduction and mixing in which molecules in a fluid (air, water, or fat) move from a warmer area to a cooler area

    d. Energy is transferred by waves of heat or light striking the food

8. A liquid, thickened with a starch, will begin to thicken gradually (depending on what starch was used) over what temperature range?

    a. 100°F to 130°F

    b. 135°F to 150°F

    c. 150° F to 212°F

    d. 212°F to 250°F

9. Which statement is *false* regarding deep fat frying?
   a. Deep-frying in a saucepan is equally as effective as using a commercial fryer.
   b. Foods deep fried together should be a similar size and shape.
   c. Delicately flavored foods should be fried separately from foods with strong flavors.
   d. Vegetable oils are the most common type of lipid used for deep frying.

10. Which of the following *two* lipids are considered flavorful fats for frying, but inferior because they cannot be kept at high cooking temperatures for long periods of time?
   a. Peanut oil
   b. Beef fat
   c. Soybean oil
   d. Pork fat (lard)

11. Which of the following is *not* important to consider when choosing the appropriate type of fat in which to fry foods?
   a. Availability
   b. Smoke point
   c. Resistance to chemical breakdown
   d. Flavor

12. Which of the following practices related to properly maintaining deep-fryer fat to maximize its lifespan is relatively unimportant?
   a. Strain and skim the fat regularly to remove food particles
   b. Avoid salting the food over the fat
   c. Store it in an airtight container
   d. Avoid exposing the fat to excessive moisture (water)

10E. Short Answer

Provide a short response that correctly answers each of the requests below.

1. List the four (4) major differences between braising and stewing

   Braising   ***Stewing***

   a. _____   a. _____

   b. _____   b. _____

   c. _____   c. _____

   d. _____   d. _____

2. List two recommendations on how to properly steam a food product.

   a. _____

   b. _____

3. In six (6) steps explain how to properly sauté a chicken breast.

   a._____

   b. _____

c. _____

d. _____

e. _____

f. _____

4. Describe the seven (7) steps necessary for correct poaching of a food item.

a. _____

b. _____

c. _____

d. _____

e. _____

f. _____

g. _____

10F. Matching

Match each of the cooking methods in List A with the appropriate temperature in List B. Each choice in List B may be used only once.

List A

_____ 1. boiling

_____ 2. broiling

_____ 3. simmering

_____ 4. poaching

_____ 5. steaming

_____ 6. deep-fat frying

List B

a. 160° F/71°C to 180° F/82°C

b. 185°F/85°C to 205° F/96°C

c. up to 2000° F/1093°C

d. 212°F/100°C or higher (at sea level)

e. 212° F/100°C (at sea level)

f. 212° F/100°C to 220° F/104°C

g. 325° F/160°C to 375° F/190°C

10G. Chapter Review

For each statement below circle either True or False to indicate the correct answer. If an answer is false, then explain why.

1. A microwave oven can be considered an acceptable replacement for traditional ovens.
   True   False

2. Sautéing is an example of the conduction heat transfer method.
   True   False

3. A wood-fired grill is an example of the convection heat transfer method.
   True   False

4. Considering proper food safety guidelines, lasagna should be baked to an internal temperature of 160°F/71°C.

   True   False

5. Most proteins complete coagulation at 160°F/71°C to 185°F/85°C.

   True   False

6. In broiling the heat source comes from below the cooking surface.

   True   False

7. Deep frying is an example of a combination cooking method.

   True   False

8. When creating crosshatch markings on a grilled steak, the meat should be rotated 90 degrees from the original position on the grill.

   True   False

9. Stir-frying is a variation in technique to sautéing except it utilizes more fat in the process.

   True   False

10. A court bouillon should be used when steaming foods in order to prevent flavor loss.

    True   False

11. Blanching means to partially cook a food product in a boiling liquid or hot fat.

    True   False

12. The process of stewing generally takes less time than braising because the food products are cut into smaller pieces.

    True   False

13. Cooking destroys harmful microorganisms and makes food easier to ingest and digest.

    True   False

14. The internal temperature of a roasted chicken breast should be 180°F/82°C.

    True   False

15. The swimming method is best to use when large quantities of food need frying.

    True   False

16. Pastas are some of the few starches that can hold up to the rapid convection movement of boiling.

    True   False

17. Wire frying baskets should be filled with foods to be fried while hanging over the fat, as to prevent the possibility of dripping fat or water on the floor or work surfaces.

    True   False

18. Grilled double lamb chops should be cooked to an internal temperature of 145°F.

    True   False

10H. Putting It All Together

Provide a short response for each of the questions below. These questions are designed to help you connect the bigger concepts presented in this chapter and/or text.

1. In Chapter Five we learned about different characteristics of metals and how some are better conductors of heat than others. Using that same concept, explain why a potato would cook faster in boiling water than baking in an oven of equal or higher temperature?

2. In Chapter Three we learned about hydrogenated fats: oils that are bombarded with hydrogen atoms, therefore being transformed from a liquid to a solid. Certain frying oils have been hydrogenated; what is the benefit to using these?

3. There are two ways of determining doneness of a braised piece of meat. One requires you to think back to Chapter Two, Food Safety & Sanitation, the other clue is in this chapter. How can you tell when it's ready to serve?

11
STOCKS AND SAUCES

> ## TEST YOUR KNOWLEDGE

The practice sets provided below have been designed to test your comprehension of the information found in this chapter. It is recommended that you read this chapter completely before attempting these questions.

11A. Terminology

Fill in the blank spaces with the correct definition.

1. Stock _____

2. Sauce _____

3. White stock _____

4. Brown stock _____

5. Fish stock _____

6. Fumet _____

7. Court bouillon _____

8. Mirepoix _____

9. Cartilage _____

10. Connective tissue _____

11. Collagen _____

12. Gelatin _____

13. Degrease _____

14. Deglaze _____

15. Remouillage _____

16. Sweat _____

17. Mother or leading sauces _____

18. Small or compound sauces _____

19. Coulis _____

20. Beurre blanc and beurre rouge _____

21. White roux _____

22. Blond roux _____

23. Brown roux _____

24. Slurry _____

25. Temper _____

26. Reduction _____

27. Béchamel _____

28. Mornay sauce _____

29. Veloute _____

30. Espagnole _____

31. Demi-glace _____

32. Chasseur sauce _____

33. Jus lie _____

34. Gastrique _____

35. Hollandaise sauce _____

36. Bearnaise _____

37. Glacage _____

38. Beurre noir and beurre noisette _____

39. Maitre d'Hôtel _____

40. Pan gravy _____

41. Chutney _____

42. Essence or tea _____

43. Flavored oil _____

44. Pesto _____

11B. Stock Making Review

Stock making is a fundamental skill. The procedure for making basic stocks should be second nature to all good chefs.

List the essential ingredients and describe in a step-by-step manner the cooking procedure for white stock, brown stock and fish stock. Exact quantities are not important for this exercise. However, cooking times should be included and each step should be numbered for production purposes.

White Stock

Ingredients: *Procedure:*

_____

_____

_____

_____

_____

_____

_____

_____

Brown Stock

Ingredients: *Procedure:*

_____

_____

_____

_____

_____

_____

_____

_____

Fish Stock

Ingredients: *Procedure:*

_____

_____

_____

_____

_____

_____

_____

_____

_____

11C. Mother Sauce Review

This section reviews the makeup of the five mother sauces. In the spaces provided below fill in the name of the sauce, the thickener used and the liquid that forms the base of the sauce. For the sauces that use a roux as a thickener, please specify the type of roux used.

Mother Sauce	Thickener	Liquid
1._____	_____	_____
2._____	_____	_____
3._____	_____	_____
4._____	_____	_____
5._____	_____	_____

11D. Small Sauces

For the following small sauces identify the leading sauce that forms its base and list the main ingredients or garnish that distinguish them from the mother sauce.

Small Sauce	Mother Sauce	Ingredients Added
1. Cream sauce	_____	_____
2. Cheddar sauce	_____	_____
3. Mornay	_____	_____
4. Nantua	_____	_____
5. Soubise	_____	_____
6. Allemande	_____	_____
7. Supreme	_____	_____
8. Bercy	_____	_____
9. Cardinal	_____	_____

Small Sauce	Mother Sauce	Ingredients Added
10. Normandy		
11. Aurora		
12. Horseradish		
13. Poulette		
14. Albufera		
15. Hungarian		
16. Ivory		
17. Bordelaise		
18. Chasseur		
19. Chateaubriand		
20. Cherveuil		
21. Madeira or Port		
22. Marchand de vin		
23. Perigeaux		
24. Piquant		
25. Poivrade		
26. Robert		
27. Creole		
28. Milanaise		
29. Spanish		
30. Bearnaise		
31. Choron		
32. Foyot		
33. Grimrod		
34. Maltaise		
35. Mousseline		

11E. Short Answer

Provide a short response that correctly answers each of the questions below.

1. List the seven (7) principles of stock making.

   a._____

   b._____

   c._____

   d._____

e._____

f._____

g._____

2. Hollandaise Preparation:

In the space below list the essential ingredients and describe in a step-by-step manner the preparation of hollandaise sauce, using the classical method. Exact quantities are important for this exercise, and each step should be numbered for production purposes.

Ingredients: *Procedure:*

_____

_____

_____

_____

_____

_____

_____

_____

3. Give five (5) reasons why hollandaise sauce might separate.

a._____

b._____

c._____

d._____

e._____

4. Provide a brief description of how the following thickening agents are combined with the liquid to form a sauce.

a. Roux _____

_____

b. Cornstarch _____

_____

c. Arrowroot _____

_____

d. Beurre Manie _____

_____

e. Liaison _____

11F. Matching

Match each of the small sauces in List A with the appropriate mother sauce in List B.

List A	List B
_____ a. Cream sauce	1. Espagnole
_____ b. Maltaise	_____
_____ c. Supreme	_____
_____ d. Bordelaise	_____
_____ e. Creole	2. Hollandaise
_____ f. Mornay	_____
_____ g. Bercy	_____
_____ h. Milanaise	_____
_____ i. Choron	3. Béchamel
_____ j. Nantua	_____
_____ k. Chasseur	_____
_____ l. Bearnaise	_____
_____ m. Cardinal	4. Velouté
_____ n. Robert	_____
_____ o. Spanish	_____
	_____
	5. Tomato
	_____
	_____
	_____

11G. Chapter Review

For each statement below, circle either True or False to indicate the correct answer. If an answer is false, then explain why.

1. More roux is needed to thicken dark sauces than to thicken light sauces.

   True   False

2. To avoid lumps in sauces, add hot stock to hot roux.

   True   False

3. After adding a liaison to a sauce, simmer for five minutes.

   True   False

4. Nappe is a term used to describe the consistency of sauce.

   True   False

5. The combination of water and cornstarch is called slurry.
    True   False

6. A velouté is a roux thickened sauce.
    True   False

7. The definition of tempering is the gradual lowering of the temperature of a hot liquid by adding a cold liquid.
    True   False

8. A reduction method is sometimes used to thicken sauces.
    True   False

9. Compound sauces come from small sauces.
    True   False

10. Fish stock needs to simmer for one hour in order to extract flavor from the ingredients.
    True   False

11. LaVarenne is credited with developing the modern system for classifying hundreds of classical sauces.
    True   False

12. Escoffier simplified Carême's extravagant list of sauces in the 19th century.
    True   False

13. Commercial bases and bouillon cubes or granules are all labor-saving convenience ingredients available to chefs.
    True   False

14. A court bouillon is derived from a nage.
    True   False

15. A chinoise mousselline is the most appropriate piece of equipment through which to strain a finished stock.
    True   False

16. Fond is the French word for stock or base.
    True   False

17. The quality of a stock is judged by its body, flavor, clarity and color.
    True   False

18. The bones from young animals, such as mature beef bones, are the best source of collagen proteins that can enhance the body of a stock.
    True   False

19. Vegetable stocks have the same body as meat stocks.
    True   False

11H. Putting It All Together

Provide a short response for each of the questions below. These questions are designed to help you connect the bigger concepts presented in this chapter and/or text.

1.   Referring back to Chapter sTwo and Three, how can the layer of fat that solidifies on top of a chilled container of stock help to preserve it?

2.   This chapter talks about cooling and storing stock properly, however, gives few details in terms of the proper procedure. Refer to Chapter Two to create a description of the process that should be followed including size and shape of the containers and how quickly the stock should cool to a particular temperature. Also make note of how a cooling wand may be used to speed up the cooling process.

12

SOUPS

➢ **TEST YOUR KNOWLEDGE**

The practice sets below have been designed to test your comprehension of the information found in this chapter. It is recommended that you read this chapter completely before attempting these questions.

12A. Terminology

Fill in the blank spaces with the correct definition.

1. Broth _____

2. Consommé _____

3. Cream soup _____

4. Purée soup _____

5. Bisque _____

6. Chowder_____

7. Cold soup _____

8. Garnish_____

9. Tomato concassée _____

10. Clearmeat or clarification _____

11. Onion brûlée _____

12. Raft_____

13. Render _____

12B. Short Answer

1. Briefly describe the six (6) steps in the preparation of a broth.

   a._____

   b._____

   c._____

   d._____

   e._____

   f._____

2. The procedure for making consommé is time tested. List the essential ingredients and describe the eight (8) steps necessary for the production of consommé.

   a._____

   b._____

   c._____

   d._____

   e._____

   f._____

   g._____

   h._____

3. Provide a reason for each of the following problems with consommé preparation.

   a. Cloudy_____

   b. Greasy_____

   c. Lacks flavor_____

   d. Lacks color_____

4. Describe the four (4) steps that can be taken to correct a poorly clarified consommé.

   a._____

   b._____

   c._____

   d._____

5. List the essential ingredients of a cream soup and describe the seven (7) essential steps for its preparation.

   a._____

   b._____

      1._____

      2._____

      3._____

4. _____

5. _____

6. _____

7. _____

6. List four (4) steps that can be taken to prevent cream from curdling when it is added to cream soups.

    a. _____

    b. _____

    c. _____

    d. _____

12C. Soup Review

1. List the seven (7) common categories of soups and provide two (2) examples of each type. In addition, suggest an appropriate garnish for each soup.

Soup	*Example*	*Garnish*
a. _____	_____	_ _____
b. _____	_____	_ _____
c. _____	_____	_ _____
d. _____	_____	_ _____
e. _____	_____	_ _____
f. _____	_____	_ _____
g. _____	_____	_ _____

2. Compare and contrast the following soups. Explain what they have in common and what makes them different from each other.

a. Beef broth   Beef Consommé

    _____

    _____

    _____

b. Cream of mushroom   Lentil soup

    _____

    _____

    _____

c. Gazpacho   Cold Consommé

    _____

    _____

    _____

12D. Chapter Review

For each statement below, circle either True or False to indicate the correct answer. If an answer is false, then explain why.

1. A purée soup is usually chunkier than a cream soup.

   True   False

2. A cream soup is always finished with milk or cream.

   True   False

3. Consommé should be stirred after the clearmeat is added.

   True   False

4. A broth is a consommé with vegetables added to it.

   True   False

5. Cream soups are thickened with a purée of vegetables which have been cooked in a stock.

   True   False

6. Bisques are generally made from shellfish, thickened with a roux and prepared using combination of cream and purée soup procedures.

   True   False

7. Cold soups should be served at room temperature.

   True   False

8. Once a consommé is clouded it should be discarded.

   True   False

9. Cold soups need less seasoning than hot soups.

   True   False

10. A roux can be used as a thickener for cold soups.

    True   False

11. If three gallons of fresh peach and yogurt soup (recipe 12.9) was produced, the amount of peaches needed in the recipe would be 24 pounds.

    True   False

12. All chowders are hearty soups, contain milk or cream and have chunks of the main ingredients.

    True   False

13. To control bacterial growth in cold soups—whether they've been cooked or not—they should be made in small batches, stored and served at or below 40°F/4°C.

    True   False

14. One of the biggest food safety challenges involved in cooling and reheating thick soups is that they scorch easily.

    True   False

15. In the recipe for Beef Consommé (recipe 12.3) the egg whites, ground beef, mirepoix, seasonings and tomato product serve as the clearmeat ingredients.

    True   False

16. If only three quarts of Beef Broth (recipe 12.1) are to be produced, the conversion factor would be 0.425.

    True   False

17. Puréeing large quantities of soup would be most efficiently achieved by using a vertical chopper mixer.

    True   False

18. Adding a final liaison to a soup as a thickening agent is not appropriate if the soup will need to be brought back to a boil before serving.

    True   False

19. Vichyssoise (recipe 12.8) contains two ingredients that are considered potentially hazardous food products.

    True   False

20. A chowder is an example of a soup that contains so many hearty, eye appealing chunks that it may not need to be garnished at service.

    True   False

12E. Putting It All Together

Provide a short response for each of the questions below. These questions are designed to help you connect the bigger concepts presented in this chapter and/or text.

1. Refer back to Chapter Two, Food Safety and Sanitation: What are the potential food safety issues associated with uncooked cold soups?

2. This chapter provides information about both cold and hot soups. Referring back to Chapter Seven, what is the rule about seasoning each type of soup?

3. Referring back to the principles you learned in Chapter Four, Menus and Recipes, and using two of the soup recipes found in this chapter, which would be more economical to produce, recipe 12.30, Artichoke Soup or recipe 12.32, Chilled Cherry Soup? Provide some specifics when explaining your answer.

13

PRINCIPLES
OF MEAT COOKERY

➢ **TEST YOUR KNOWLEDGE**

The practice sets provided below have been designed to test your comprehension of the information found in this chapter. It is recommended that you read this chapter completely before attempting these questions.

13A. Terminology

1. Primal cuts _____

2. Subprimal cuts _____

3. Fabricated cuts _____

4. Marbling_____

5. Subcutaneous fat _____

6. Collagen_____

7. To butcher _____

8. Dress _____

9. Fabricate_____

10. Carve _____

11. Quality grades _____

12. Yield grades _____

13. Vacuum packaging_____

14. Portion control _____

15. Freezer burn _____

16. Carryover cooking _____

17. Fond _____

18. Cutlet _____

19. Scallop _____

20. Émincé _____

21. Paillard _____

22. Medallion _____

23. Mignonette _____

24. Noisette _____

25. Chop _____

26. Brown stew _____

27. White stew/fricassee _____

28. Blanquette _____

13B. Fill in the Blank

Fill in the blank with the response that correctly completes the statement.

1. Dry heat cooking methods are best used for _____ _____ cuts of meat.

2. Once the meat is added to the sauce for stewing, the dish is cooked at a _____ temperature for a _____ time.

3. Very rare meats should feel _____ and have a _____ _____ color; however, well-cooked meats should feel _____ and have a _____-_____ color.

4. Sauté items are sometimes _____ in _____ before being placed in the hot pan.

5. Some _____ cooking occurs when a roast item is removed from the oven. Allowing the meat to rest will help the meat to _____ more _____.

6. Covering the exterior surface with fat before cooking is called _____ , and inserting strips of fat into the meat is called _____ .

7. Marinating meats adds a distinctive _____ and breaks down the to help tenderize the meat.

8. Ionizing radiation kills significant amounts of _____ , _____ and in meat.

9. Dry heat cooking methods are not recommended for _____ cuts of meat, or those high in connective tissue.

10. When carving roasted meats, it is important to carve _____ the grain.

13C. Matching

Match each of the stews in List A with the appropriate description in List B. Each choice in List B can only be used once.

List A	List B
_____ 1. Ragoût	a. A spicy ragoût of ground or diced meat with vegetables, peppers and sometimes beans.
_____ 2. Goulash	b. A white stew usually made with white meat and garnished with onions and mushrooms.
_____ 3. Blanquette	c. A brown ragoût made with root vegetables and lamb.
_____ 4. Fricassee	d. A general term that refers to stews.
_____ 5. Navarin	e. A Hungarian beef stew made with onions and paprika and garnished with potatoes.
	f. A white stew in which the meat is blanched and added to the sauce to finish the cooking process. The stew is finished with a liaison of cream and egg yolks.

13D. Cooking Methods

Briefly describe each of the following methods of cooking and provide an example of a cut of meat used in each method.

Cooking method	Description	Example
Grilling	_____	_____
Roasting	_____	_____
Sautéing	_____	_____
Pan-frying	_____	_____
Simmering	_____	_____
Braising	_____	_____
Stewing	_____	_____

13E. Chapter Review

For each statement below, circle True or False to indicate the correct answer. If an answer is false, then explain why.

1. Fresh meats should be stored at 35°F to 40°F.

   True   False

2. "Green meats" are meats that are allowed to turn moldy.

   True   False

3. Braising and stewing are combination cooking methods.

   True   False

4. The USDA stamp on whole carcasses of meat does not ensure their quality or tenderness.

   True   False

5. USDA Choice meat is used in the finest restaurants and hotels.

   True   False

6. Yield grades are used for beef, lamb and pork.

   True   False

7. Wet aging occurs in a vacuum package.

   True   False

8. During dry aging the meat may develop mold, which adds to the flavor of the meat.

   True   False

9. Under the correct conditions, vacuum-packed meat can be held for 2 to 3 weeks.

   True   False

10. Still air freezing is the most common method of freezing meats in food service facilities.

    True   False

11. The fat cap on a piece of meat adds tenderness and is a principal factor in determining meat quality.

    True   False

12. Broiled chicken livers wrapped in bacon is an example of a barded meat product.

    True   False

13. In order to account for carryover cooking, an 8 pound roasted pork loin should be removed from the oven when the internal temperature is approximately 140°F.

    True   False

14. Roasting meats at a lower temperature, such as 300°F to 325°F, may result in better yield and a slightly juicier final product.

    True   False

15. All meat produced for public consumption in the United States is subject to USDA inspection, which indicates a meat's quality and tenderness.

    True   False

15. Once the Sautéed Veal Scallops with White Wine Lemon Sauce (recipe 13.3) is converted to yield 10 portions, the amount of veal needed is 3.76 pounds.

    True   False

17. In order to account for carryover cooking, a steamship round should be removed from the oven when the internal temperature is approximately 145°F.

    True   False

18. The conversion factor to change the New England Boiled Dinner (recipe 13.5) from 12 portions to 30 is 0.40.

    True   False

19. A whole stuffed chicken would benefit from being larded.

    True   False

20. Scallops wrapped in bacon is an example of the barding technique.

    True   False

13F. Putting It All Together

Provide a short response for each of the questions below. These questions are designed to help you connect the bigger concepts presented in this chapter and/or text.

1. In Chapter Two, Food Safety and Sanitation, you learned about the importance for chefs to adhere to the time/temperature principles within the HACCP system. Relate the same principles to primary suppliers of raw meats who undergo the grading process. What would a chef's job be like if the primary suppliers don't follow HACCP closely?

2. This chapter references a safety alert when it comes to proper internal temperatures to which various meats should be cooked in order to be considered safe for consumption. Based on your knowledge of food safety and sanitation, why is it not recommended that beef or fresh pork be served rare?

3. Without turning back to Chapter Eleven, Stocks and Sauces, to reference this, and without referencing the recipe, briefly describe the *au jus* that is served with the roast prime rib of beef in recipe 13.2.

14

BEEF

> ## TEST YOUR KNOWLEDGE

The practice sets provided below have been designed to test your comprehension of the information found in this chapter. It is recommended that you read this chapter completely before attempting these questions.

14A. Terminology

For each primal cut of beef listed below, describe the following:

    a.   % of carcass weight

    b.   Bone structure

    c.   Muscle structure

    d.   Cooking processes applied

1. Chuck _____

    a. _____

    b. _____

    c. _____

    d. _____

2. Brisket and Shank _____

    a. _____

    b. _____

    c. _____

    d. _____

3. Rib _____

    a. _____

    b. _____

    c. _____

    d. _____

4. Short plate _____

    a. _____

    b. _____

    c. _____

    d. _____

5. Short loin _____

    a. _____

    b. _____

    c. _____

    d. _____

6. Sirloin _____

    a. _____

    b. _____

    c. _____

    d. _____

7. Flank _____

    a. _____

    b. _____

    c. _____

    d. _____

8. Round _____

    a. _____

    b. _____

    c. _____

    d. _____

14B. Primal Cuts of Beef

Identify the primal cuts of beef indicated in the following diagram and write their names in the spaces provided.

1._____

2._____

3._____

4._____

5._____

6._____

7._____

8._____

14C. Cuts from the Round

Name the five (5) subprimal/fabricated cuts from the round and name the most appropriate cooking process or use for each cut.

Subprimal/Fabricated Cut	Cooking Process/Use
1._____	_____
2._____	_____
3._____	_____
4._____	_____
5._____	_____

14D. Cuts of Beef and Applied Cooking Methods

Name the cooking method applied to the main ingredient in each of the following beef dishes. Also, identify a subprimal and a primal cut of meat from which the main ingredient is taken.

Name of Dish	Cooking Method	Subprimal/ Fabricated Cut	Primal Cut
1. Pot Roast	_____	_____	_____
2. Tamales / hash	_____	_____	_____
3. Entrecotes Bordelaise	_____	_____	_____
4. Hamburgers / meatloaf	_____	_____	_____

Name of Dish	Cooking Method	Subprimal/ Fabricated Cut	Primal Cut
5. Stuffed flank steak	_____	_____	_____
6. New England boiled dinner	_____	_____	_____
7. Tournedos Rossini	_____	_____	_____
8. London broil	_____	_____	_____
9. Beef Wellington	_____	_____	_____
10. Roast beef	_____	_____	_____
11. Beef roulade	_____	_____	_____
12. Beef fajitas	_____	_____	_____
13. Beef stew	_____	_____	_____
14. Chili con carne	_____	_____	_____
15. Minute steak	_____	_____	_____

14E. Multiple Choice

For each request below, choose the one correct response.

1. The outside round and the eye of the round together are called the:
   a. Top round
   b. Bottom round
   c. Steamship round
   d. Primal round

2. A carcass of beef weighs:
   a. Between 600 and 950 pounds
   b. Up to 1,000 pounds
   c. From 500 to more than 800 pounds
   d. Between 400 and 600 pounds

3. The three fabricated cuts from the tenderloin are:
   a. Porterhouse steak, tournedos and chateaubriand
   b. Butt tenderloin, filet mignon and tournedos
   c. Chateaubriand, short loin and loin eye
   d. Tournedos, chateaubriand and filet mignon

4. "Butterflying" is a preparation technique that:
   a. Makes the cut of meat thinner
   b. Makes the meat more tender
   c. Improves the flavor of the meat
   d. Improves the appearance of the meat

5. Pastrami and corned beef are both flavored, preserved meats that come from:

    a. Brisket

    b. Chuck

    c. Round

    d. Flank

6. If the primal chuck accounts for approximately 28% of carcass weight, how much would the chuck weigh on a 728 pound carcass?

    a. 524.16 pounds

    b. 333.2 pounds

    c. 203.84 pounds

    d. 178.46 pounds

7. Put the following steaks in order based on the size portion of tenderloin they contain (1=largest portion, 3=smallest portion)

    _____ a. T-bone

    _____ b. Porterhouse

    _____ c. Club

14F. Matching I

Match each of the primal cuts in List A with the appropriate description in List B. Each choice in List B can only be used once.

List A

_____ 1. Rib

_____ 2. Chuck

_____ 3. Short Loin

_____ 4. Flank

List B

a. Produces the boneless strip loin, which can be roasted whole or cut into steaks.

b. This cut is located in the hindquarter between the short loin and the round.

c. The eye of this cut is well exercised, quite tender and contains large quantities of marbling. It is suitable for roasting.

d. The animal constantly uses the muscle in this primal cut, therefore it is tough, contains high levels of connective tissue and is very flavorful.

e. This primal cut produces the hanging tenderloin which is very tender and can be cooked by any method.

14G. Matching II

Match the following cooking processes in List A with the appropriate cut of meat in List B. Each cooking process matches *only* three (3) cuts of meat.

List A	**List B**
1. Grill	a. ____ Shank
	b. ____ Porterhouse
2. Roast	c. ____ Rib
	d. ____ Top round
3. Braise	e. ____ Strip loin
	f. ____ Short ribs
	g. ____ Skirt steak
	h. ____ Knuckle
	i. ____ Bottom round

14H. Chapter Review

For each statement below, circle either True or False to indicate the correct answer. If an answer is false, then explain why.

1. The hanging tenderloin is part of the flank.

   True   False

2. The meat from the chuck is less flavorful than meat from the tenderloin.

   True   False

3. A porterhouse steak is fabricated from the tenderloin.

   True   False

4. Prime rib of beef refers to the quality USDA grade.

   True   False

5. The subprimal and fabricated cuts from the short loin are the most tender and expensive cuts of beef.

   True   False

6. The subprimal and fabricated cuts from the sirloin are not as tender as those from the strip loin.

   True   False

7. The short loin can be cut across to produce porterhouse, T-bone and club steaks.

   True   False

8. Pastrami is made from the meat in the short plate.

   True   False

9. London Broil comes from the flank.

   True   False

10. Foreshanks and hindshanks are flavorful cuts of beef that are best ground and used for hamburgers.

   True   False

11. In order to prepare twenty 4-ounce appetizer portions of Beef Bourguignon (recipe 14.17), 8 pounds of beef chuck are required.

   True   False

12. To produce a Maillard reaction and achieve good browning on the outside of a steak, a chef needs to use a dry cooking technique with temperatures greater than 300°F.

   True   False

14I.   Putting It All Together

Provide a short response for each of the questions below. These questions are designed to help you connect the bigger concepts presented in this chapter and/or text.

1. In reviewing the primal cuts of beef, this chapter reviewed the rib. It is common practice when preparing to roast a prime rib to remove the fat cap, season the meat beneath, then return the fat cap to the meat. In review of Chapter Thirteen, Principles of Meat Cookery, what does the chef have to do to the fat cap in order to keep it in touch with the meat during the roasting process? Secondly, what is the technique called when you wrap the outside of a piece of meat in fat in order to add moisture and flavor during the cooking process?

2. Based on what you've learned in this chapter and your knowledge of menus and recipe costing from Chapter Four would it make good financial sense to braise a strip loin? Explain your answer.

15

VEAL

➢ **TEST YOUR KNOWLEDGE**

The practice sets provided below have been designed to test your comprehension of the information found in this chapter. It is recommended that you read this chapter completely before attempting these questions.

15A. Terminology

Fill in the blank spaces with the correct definition.

1. Foresaddle _____

2. Hindsaddle _____

3. Back _____

4. Veal side _____

5. Sweetbreads _____

6. Calves' liver _____

7. Kidneys _____

8. Veal scallops _____

9. Émincé _____

15B. Primal Cuts Identification

For each primal cut of Veal listed below, describe the following:

    a. % of carcass weight

    b. Bone structure

    c. Muscle structure

    d. Cooking processes applied

1. Shoulder _____

    a. _____

    b. _____

    c. _____

    d. _____

2. Foreshank and breast _____

    a. _____

    b. _____

    c. _____

    d. _____

3. Rib _____

    a. _____

    b. _____

    c. _____

    d. _____

4. Loin _____

    a. _____

    b. _____

    c. _____

    d. _____

5. Leg _____

    a. _____

    b. _____

    c. _____

    d. _____

15C. Primal Cuts of Veal

Identify the primal cuts of veal indicated in the following diagram and write their names in the spaces provided.

1._____

2._____

3._____

4._____

5._____

15D. Cuts of Veal and Applied Cooking Methods

Name the cooking method applied to the main ingredient in each of the following dishes. Also, identify a subprimal and a primal cut of meat from which the main ingredient is taken.

Name of Dish	Cooking Method	Subprimal/ Fabricated Cut	Primal Cut
1. Blanquette/fricassee	_____	_____	_____
2. Veal rib eye/ Marchand de vin	_____	_____	_____
3. Stuffed veal breast	_____	_____	_____
4. Veal chop with mushroom sauce	_____	_____	_____
5. Blanquette	_____	_____	_____
6. Veal sweetbreads	_____	_____	_____
7. Kidney pie	_____	_____	_____
8. Veal patties	_____	_____	_____
9. Meatballs	_____	_____	_____
10. Tenderloin	_____	_____	_____
11. Veal Marsala	_____	_____	_____
12. Calves' liver	_____	_____	_____
13. Osso Buco	_____	_____	_____
14. Veal scallopini	_____	_____	_____
15. Veal broth	_____	_____	_____

15E. Short Answer

Provide a short response that correctly answers the following requests.

1. Briefly describe the eight (8) basic steps to be followed when boning a leg of veal, beginning with:

   a. <u>Remove the shank</u>

   b. _____

   c. _____

   d. _____

   e. _____

   f. _____

   g. _____

   h. _____

2. Name the six (6) muscles in the leg of veal.

   a. _____

   b. _____

   c. _____

   d. _____

   e. _____

   f. _____

3. Name the three (3) subprimal cuts from the rib and three (3) from the loin and provide a menu example of each cut.

Primal Cut	Subprimal/Fabricated Cut	Menu Example
a. Rib	_____	_____
b. Rib	_____	_____
c. Rib	_____	_____
d. Loin	_____	_____
e. Loin	_____	_____
f. Loin	_____	_____

4. Compare and contrast formula-fed veal with free-range veal. Discuss the advantages and disadvantages of each.

   _____

   _____

   _____

   _____

15F. Matching I

Match the primal cuts in List A with the appropriate definitions in List B. Each choice in List B can be used only once.

List A

_____1. Leg

_____2. Shoulder

_____3. Foreshank

_____4. Loin

_____5. Rib

List B

a. A cut of veal similar to the chuck in beef

b. A primal cut of veal which is located just below the shoulder and rib section in the front of the carcass

c. The primal cut that produces the short tenderloin and breast

d. The bones in this cut are still soft, due to the immaturity of the animal

e. The primal cut that yields the most tender meat

f. Includes part of the backbone, tail bone, hip bone, aitch bone, round bone and shank.

15G. Matching II

Match each of the following cooking processes with the appropriate cut of meat. Each process matches *only* two (2) cuts of meat.

1. Sauté

2. Roast

3. Braise

a. _____ Split veal rack

b. _____ Calves' liver

c. _____ Veal shank

d. _____ Whole veal loin

e. _____ Veal émincé

f. _____ Veal breast

15H. Chapter Review

For each statement below, circle either True or False to indicate the correct answer. If an answer is false, then explain why.

1. Veal scallops are taken from large pieces of veal and are cut on the bias, across the grain of the meat.

   True   False

2. Veal flesh begins to change color when the animal consumes iron in its food.

   True   False

3. Sweetbreads are pressed to remove the impurities.

   True   False

4. Veal émincé are cut with the grain, from small pieces of meat.

   True   False

5. Veal scallops are pounded in order to make them more tender.

    True   False

6. Sweetbreads become larger as the animal ages.

    True   False

7. Veal liver has a more delicate flavor than beef liver.

    True   False

8. The hindshank and foreshank of veal are prepared and cooked in the same manner.

    True   False

9. The feed given to calves during their short lifespan is formulated to keep their muscles from toughening.

    True   False

10. While it is a more humane method of producing veal, less free range veal is raised and therefore it is more expensive.

    True   False

11. On a veal carcass weighing 175 pounds, the primal leg would represent 73.5 pounds.

    True   False

12. A good use for veal shoulder meat would be butterflied for Veal Marsala.

    True   False

13. Veal Cordon Bleu is made by rolling ham and Swiss cheese within a veal cutlet, coating it using a standard breading procedure, and then deep-fat frying.

    True   False

14. Osso Buco is classically braised in order to further concentrate the flavors of the foreshank.

    True   False

15. The Stuffed Veal Breast (recipe 15.3) should be cooked to an internal temperature of 165°F/74°C or until tender.

    True   False

15I.   Putting It All Together

Provide a short response for each of the questions below. These questions are designed to help you connect the bigger concepts presented in this chapter and/or text.

1. Based on what you now know about veal, why are most animals male?

2. Considering what you learned about high amounts of fat and cholesterol in the diet in Chapter Three, Nutrition, explain why including veal in place of other more mature beef cuts would be advantageous.

3. Based on your newfound knowledge of veal, its age and level of activity in its short life, how long do you think a tough cut of veal like the leg or bottom round has to stew or braise compared to the same cut of beef?

16

LAMB

> ## TEST YOUR KNOWLEDGE

The practice sets provided below have been designed to test your comprehension of the information found in this chapter. It is recommended that you read the chapter completely before attempting these questions.

16A. Terminology

Fill in the blank spaces with the correct definition.

1. Foresaddle _____

2. Hindsaddle _____

3. Back _____

4. Bracelet _____

16B. Primal Cuts Identification

For each primal cut of lamb listed below describe the following:

    a. % of carcass weight

    b. Bone structure

    c. Muscle structure

    d. Cooking processes applied

1. Shoulder _____

    a._____

    b. _____

    c._____

    d. _____

2. Breast _____

     a. _____

     b. _____

     c. _____

     d. _____

3. Rack _____

     a. _____

     b. _____

     c. _____

     d. _____

4. Loin _____

     a. _____

     b. _____

     c. _____

     d. _____

5. Leg _____

     a. _____

     b. _____

     c. _____

     d. _____

16C. Primal Cuts of Lamb

Identify the primal cuts of lamb indicated in the following diagram and write their names in the spaces provided below.

1. _____

2. _____

3. _____

4. _____

5. _____

16D. Subprimal or Fabricated Cuts

For each primal cut in section 16C., name two subprimal cuts and an appropriate cooking method for each of these cuts.

	Primal Cut	*Subprimal/Fabricated Cut*	*Cooking methods*
1.		a._____	_____
		b. _____	_____
2.		a._____	_____
		b. _____	_____
3.		a._____	_____
		b. _____	_____
4.		a._____	_____
		b. _____	_____
5.		a._____	_____
		b. _____	_____

16E. Cuts of Lamb and Applied Cooking Methods

Name the cooking method applied to the main ingredient in each of the following lamb dishes. Also, identify a subprimal and a primal cut of meat from which the main ingredient is taken.

Name of Dish	*Cooking Method*	*Subprimal/ Fabricated Cut*	*Primal Cut*
1. Lamb Kebabs	_____	_____	_____
2. Lamb curry	_____	_____	_____
3. Noisettes of lamb with roasted garlic sauce	_____	_____	_____
4. Lamb stew	_____	_____	_____
5. Broiled lamb with mustard and hazelnut crust	_____	_____	_____
6. Lamb breast stuffed with mushrooms	_____	_____	_____
7. Rack of lamb	_____	_____	_____
8. Leg of lamb	_____	_____	_____

16F. Short Answer

Provide a short response that correctly answers the following requests.

1. Briefly describe the six (6) basic steps to follow when frenching a rack of lamb.

   a._____

   b. _____

   c._____

   d. _____

   e._____

   f._____

2. Briefly describe the four (4) basic steps to follow when trimming a leg of lamb for roasting/grilling.

   a._____

   b. _____

   c._____

   d. _____

3. Briefly describe the eight (8) basic steps to be followed when preparing a loin of lamb for roasting.

   a._____

   b. _____

   c._____

   d. _____

   e._____

   f._____

   g. _____

   h. _____

16G. Chapter Review

For each statement below, circle either True or False to indicate the correct answer. If an answer is false, then explain why.

1. The lamb carcass is classified into two parts: the hindquarter and the forequarter.

   True   False

2. The term "spring lamb" applies to animals that are born between February and May.

   True   False

3. The primal cuts of both veal and lamb are broken down into bilateral halves.

   True   False

4. The primal leg of lamb is rarely left whole.

   True   False

5. The leg of lamb can be broken down to produce steaks.

   True   False

6. The chine bone, part of the backbone, runs through the loin of lamb.

   True   False

7. The "fell" refers to the thin layer of connective tissue on the outside of the loin of lamb.

   True   False

8. Denver ribs are ribs that are cut from the breast of lamb.

   True   False

9. Frenching is a process that can only be performed on lamb racks or chops.

   True   False

10. A conversion factor of 1.25 would yield 20 portions of Irish Lamb Stew (recipe 16.15).

    True   False

11. The tender rib eye comes from the loin of the lamb.

    True   False

12. To produce a more economical stew, the chef should use the shank end of the lamb leg in the Lamb in Indian Coconut Curry Sauce (recipe 16.17).

    True   False

13. Lamb patties with mint should be cooked to an internal temperature of 160°F/71°C.

    True   False

14. Because most lamb is slaughtered under the age of 1 year, its meat is tender and can be prepared by almost any cooking method.

    True   False

16H. Putting It All Together

Provide a short response for each of the questions below. These questions are designed to help you connect the bigger concepts presented in this chapter and/or text.

1. What can you surmise about domestic lamb based on the comparative sidebar that discussed domestic versus imported lamb? What does it mean for a chef planning a menu for his/her restaurant?

2. What benefits can one gain nutritionally from adding more lamb to the diet than other meats?

3. How does Chapter Four, Menus and Recipes; Chapter Nine, Mise en Place; and Chapter Six, Knife Skills; help a chef to decide whether to order a whole hotel rack to break down into its component parts as opposed to buying it already dismantled?

17

PORK

> ## ➤ TEST YOUR KNOWLEDGE

The practice sets provided below have been designed to test your comprehension of the information found in this chapter. It is recommended that you read this chapter completely before attempting these questions.

17A. Terminology

For each primal cut of pork listed below describe the following:

    a.   % of carcass weight

    b.   Bone structure

    c.   Muscle structure

    d.   Cooking processes applied

1. Boston Butt _____

    a._____

    b. _____

    c._____

    d. _____

2. Loin _____

    a._____

    b. _____

    c._____

    d. _____

3. Fresh Ham _____

    a. _____

    b. _____

    c. _____

    d. _____

4. Belly _____

    a. _____

    b. _____

    c. _____

    d. _____

    a. _____

    b. _____

    c. _____

    d. _____

17B. Primal Cuts of Pork

Identify the primal cuts of pork indicated in the following diagram and write their names in the spaces provided below.

1. _____

2. _____

3. _____

4. _____

5. _____

17C. Subprimal or Fabricated Cuts

For each primal name a subprimal cut and an appropriate cooking method. Indicate in the appropriate column(s) whether these cuts are usually smoked or fresh (or both).

Example:

Fresh ham	Hock	Braise	X	
Primal Cut	**Subprimal/ Fabricated Cut**	**Cooking Methods**	**Cured & Smoked**	**Fresh**
1. _____	_____	_____	_____	_____
2. _____	_____	_____	_____	_____
3. _____	_____	_____	_____	_____
4. _____	_____	_____	_____	_____
5. _____	_____	_____	_____	_____

17D. Cuts of Pork and Applied Cooking Methods

Name the cooking method applied to the main ingredient in each of the following recipes. Also identify a subprimal and a primal cut of meat from which the main ingredient is taken.

Name of Dish	Cooking Method	Subprimal/ Fabricated Cut	Primal Cut
1. Roast pork with apricots & almonds	_____	_____	_____
2. Choucroute	_____	_____	_____
3. Pork tenderloin	_____	_____	_____
4. Pork chops	_____	_____	_____
5. Spare ribs	_____	_____	_____
6. Smoked picnic shoulder	_____	_____	_____
7. Breakfast meat	_____	_____	_____

17E. Short Answer

Provide a short response that correctly answers the following requests.

1. Briefly describe the three (3) basic steps to be followed when boning a pork loin.

   a._____

   b_____

   c._____

2. Name six (6) fabricated cuts that are most often smoked and cured.

a._____

b. _____

c._____

d. _____

e._____

f. _____

17F. Matching

Match the primal cuts in List A with the appropriate definitions in List B. Each choice in List B can only be used once.

List A

_____1. Boston Butt

_____2. Shoulder

_____3. Belly

_____4. Loin

_____5. Fresh ham

List B

a. A primal cut that is very fatty with strips of lean meat.

b. The primal cut from which the most tender portion of pork is taken.

c. A primal cut which contains large muscles, relatively small amounts of connective tissue, may be smoked and cured or cooked fresh.

d. A primal cut with a good percentage of fat to lean meat, ideal when a solid piece of pork is required for a recipe.

e. A single, very tender eye muscle that can be braised/roasted/sautéed.

f. One of the toughest cuts of pork that has a relatively high ratio of bone to lean meat, is relatively inexpensive and is widely available.

17G. Chapter Review

For each statement below, circle either True or False to indicate the correct answer. If an answer is false, then explain why.

1. The Boston butt is located in the hindquarter.

    True   False

2. Pork is unique because the ribs and loin are considered one primal cut.

    True   False

3. The term "meat packing" originated in colonial times when pork was packed into barrels for shipment abroad.

    True   False

4. The foreshank is also known as the ham hock.

    True   False

5. Center-cut pork chops are chops that are split open to form a pocket.

   True   False

6. The belly is used to make Canadian Bacon.

   True   False

7. Backfat is the layer of fat between the skin and the lean muscle of the pork loin.

   True   False

8. Hogs are bred to produce short loins.

   True   False

9. Picnic ham is made from the hog's hind leg.

   True   False

10. The loin is the only primal cut of pork not typically smoked or cured.

    True   False

11. All pork comes from hogs slaughtered at a young age and therefore all cuts are tender enough to be cooked by any method.

    True   False

12. Spareribs are cut from the bottom portion of the primal loin.

    True   False

13. The fresh ham is the hog's hind leg.

    True   False

17H. Putting It All Together

Provide a short response for each of the questions below. These questions are designed to help you connect the bigger concepts presented in this chapter and/or text.

1. Throughout the last few chapters you've learned about the nutritional value of the various meats. Compare and contrast lamb and pork.

2. Referring to Table 17.1, Using Common Cuts of Pork, hypothesize why spareribs are usually cooked by both steaming/boiling *and* then grilling. Incorporate your knowledge of what makes a cut of meat tough or tender as well as your understanding of regional food traditions when developing your answer.

18

POULTRY

> ## TEST YOUR KNOWLEDGE

The practice sets provided below have been designed to test your comprehension of the information found in this chapter. It is recommended that you read this chapter completely before attempting these questions.

18A. Terminology

Fill in the blank spaces with the correct definition.

1. Duckling_____

2. Guinea _____

3. Squab _____

4. Dinde _____

5. Ratites _____

6. Giblets _____

7. Foie gras_____

8. Supreme _____

9. A point _____

10. Trussing _____

11. Barding_____

12. Basting _____

13. Dressing _____

14. Poulet de Bresse_____

18B. Short Answer

Provide a short response that correctly answers the following requests.

1. List five (5) important guidelines for stuffing poultry.

   a._____

   b._____

   c._____

   d._____

   e._____

2. Name one (1) similarity and one (1) major difference between poultry and red meats.

   Similarity: _____

   Difference: _____

3. Poultry is a highly perishable product and improper storage can lead to food poisoning. Including *exact temperatures and times*, discuss the guidelines for storing poultry products under the following headings.

   Storage and refrigeration: _____

   _____

   Freezing: _____

   _____

   Thawing: _____

   _____

   Reheating: _____

   _____

4. List five (5) differences between the rearing and sale of free-range chicken compared to traditionally reared chicken.

   a._____

   b._____

   c._____

   d._____

   e._____

5. What are the six (6) kinds of poultry?

   a._____

   b._____

   c._____

   d._____

   e._____

   f._____

6. For each of the following cooking methods, give a recipe example and accompaniment for the poultry item.

Cooking method	Recipe example	Accompaniment
Sauté	_____	_____
Pan-fry	_____	_____
Simmer/Poach	_____	_____
Braise/Stew	_____	_____

7. Describe five (5) ways to prevent cross-contamination when handling poultry.

a._____

b. _____

c._____

d. _____

e._____

8. Describe the six (6) steps for portioning poultry into 8 pieces.

a._____

b. _____

c._____

d. _____

e._____

f. _____

18C. Matching

Match each of the terms in List A with the appropriate letter definition in List B. Each choice in List B can only be used once.

List A

_____1. Hen/Stewing

_____2. Broiler/fryer

_____3. Roaster

_____4. Capon

_____5. Game hen

List B

a. Young tender meat, smooth skin, breastbone less flexible than a broiler's (3-5 months).

b. Mature female, less tender meat, nonflexible breastbone (over 10 months).

c. Young immature offspring of Cornish chicken, very flavorful (5-6 weeks).

d. Rich, tender dark meat with large amounts of fat, soft windpipe (6 months or less).

e. Young with soft, smooth skin, lean with flexible breastbone (13 weeks).

f. Surgically castrated male, tender meat, smooth skin, high proportion of light to dark meat, relatively high fat content (under 8 months).

18D. Fill in the Blank

Fill in the blank provided with the response that correctly completes the statement.

1.   The color difference between the legs and the wings of chicken and turkey is due to a higher concentration of the _____ called _____ in the tissue.

2.   The internal temperature of fully cooked poultry should be between _____ and _____.

3.   The most commonly used duck in food service operations is a _____. Its meat is different from chicken in two ways: the flesh is _____ and has large amounts of _____.

4.   Chicken is often marinated in a mixture of _____ or _____, salt, _____ and _____. A common example of a chicken marinade is _____ sauce.

18E. Multiple Choice

For each request below, choose one correct response.

1.   Fresh poultry should be refrigerated between:
   a.   30-35°F.
   b.   33-36°F.
   c.   32-34°F.
   d.   30-38°F.

2.   Which of the following groups does not qualify as "poultry"?
   a.   Chicken, duck, pigeon
   b.   Duck, pheasant, goose
   c.   Pigeon, guinea, chicken
   d.   Quail, duck, turkey

3.   The poultry that is sold in wholesale or retail outlets carries the USDA Grade:
   a.   A
   b.   B
   c.   C
   d.   A, B and C

4.   Due to its percentage of bone and fat to meat, a 4 pound duck will serve how many people?
   a.   1
   b.   2
   c.   3
   d.   4

18F. Chapter Review

For each statement below, circle either True or False to indicate the correct answer. If an answer is false, then explain why.

1. Poultry fat has a higher melting point than other animal fats.

   True   False

2. Duck and goose must be roasted at a higher temperature in order to render as much fat from the skin as possible.

   True   False

3. Myoglobin is a protein that stores oxygen.

   True   False

4. The longer chicken is left in a marinade, the better the flavor.

   True   False

5. Dark meat takes less time to cook than light meat.

   True   False

6. Poultry should not be frozen below 0°F/-18°C.

   True   False

7. The skin color of poultry is partly affected by the amount of sunlight to which it is exposed.

   True   False

8. Quality USDA grades do not reflect the tenderness of the poultry.

   True   False

9. Older male birds have more flavor than female birds.

   True   False

10. When foie gras is overcooked it becomes tough.

    True   False

11. A young pigeon is called a yearling.

    True   False

12. The gizzard is a term used to describe the chicken's neck.

    True   False

13. A capon is a type of pigeon.

    True   False

14. Poultry is divided into classes based upon the sex of the bird.

    True   False

15. Most ratite meat is cut from the back of birds slaughtered between 10 to 13 months.

    True   False

16. Ostrich meat is best cooked well done.

    True   False

17. Poultry should not be marinated for longer than two hours.

    True   False

18. After marinating poultry, the marinade may be stored for future use.

    True   False

19. Poultry should be cooked to an internal temperature of 160° F to 170°F.

    True   False

20. Fresh chicken should be stored refrigerated between 32° F and 34° F for up to four days.

    True   False

21. The "poultry" class includes any domesticated birds bred for eating.

    True   False

22. Although it is classified as poultry, ratite meat is a dark, cherry-red color with a flavor similar to beef but sweeter.

    True   False

23. All poultry produced for public consumption in the United States is subject to USDA inspection.

    True   False

24. The 1 ounce of clarified butter needed in the Chicken Sauté with Onions, Garlic and Basil (recipe 18.4) is equal to a volume measure of 1/8 cup.

    True   False

25. The conversion factor for 40 portions of the Apricot and Bourbon Grilled Chicken (recipe 18.11) is 10.

    True   False

26. A squab's meat is dark, tender and low in fat and therefore it would benefit from being barded before cooking.

    True   False

27. Emu and ostrich meat, even though it is classified as poultry, is a dark, cherry-red color and tastes similar to beef.

    True   False

28. Ratite meat, like chicken and most other poultry, should be cooked to an internal temperature of 165°F.

    True   False

18G. Putting It All Together

Provide a short response for each of the questions below. These questions are designed to help you connect the bigger concepts presented in this chapter and/or text.

1. Considering the yield of meat on a duck compared to a chicken, what does a chef have to take into consideration when menu planning and costing recipes?

2. After gaining a better understanding of flavors and flavorings in Chapter Seven, what might you use in a marinade for chicken? Make several suggestions for ingredients that might be used for each of the 3 main ingredients in a marinade and mention how long the meat should be marinated.

19

GAME

➤ TEST YOUR KNOWLEDGE

The practice sets provided below have been designed to test your comprehension of the information found in this chapter. It is recommended that you read this chapter completely before attempting these questions.

19A. Terminology

For each of the following game animals listed below, list the following:

    a.   Source of animal

    b.   Composition of flesh

    c.   Recommended cooking methods

1. Antelope: _____

    a._____

    b. _____

    c._____

2. Bison:_____

    a._____

    b. _____

    c._____

3. Venison:_____

    a._____

    b. _____

    c._____

4. Rabbit: _____

    a. _____

    b. _____

    c. _____

5. Wild Boar: _____

    a. _____

    b. _____

    c. _____

6. Partridge: _____ ____

    a. _____

    b. _____

    c. _____

7. Pheasant: _____

    a. _____

    b. _____

    c. _____

8. Quail: _____

    a. _____

    b. _____

    c. _____

19B. Short Answer

Provide a short response that correctly answers the following requests.

1. List three (3) uses for tougher cuts of game.

    a. _____

    b. _____

    c. _____

2. Explain the process and purpose of hanging wild game.

    Process: _____

    _____

    _____

    Purpose: _____

    _____

3. Describe the origins and composition of Beefalo.

_____

_____

_____

4. Describe the guidelines for refrigeration and freezing of game.

Refrigeration: _____

_____

Freezing: _____

_____

19C. Multiple Choice

For each request below, choose the one correct response.

1. Which of the following cannot be categorized as furred game?
   a. Antelope
   b. Pheasant
   c. Bison
   d. Rabbit

2. Due to the lean nature of game birds, they are barded and cooked:
   a. Medium
   b. Rare
   c. Medium rare
   d. Well done

3. Which one of the following is *not* a member of the deer family?
   a. Elk
   b. Bison
   c. Mule deer
   d. Moose

4. Feathered game include:
   a. Pheasant, quail, woodcock
   b. Partridge, pheasant, pigeon
   c. Turkey, lark, squab
   d. Guinea, goose, duck

5. Quail weighs approximately:
   a. 10-12 ounces
   b. 1-2 pounds
   c. 4-5 pounds
   d. 4-8 ounces

6. The most popular game bird is:
   a. Quail
   b. Partridge
   c. Pheasant
   d. Woodcock

7. If the Chili-Rubbed Venison with Caramelized Beer Sauce (recipe 19.4) was converted to yield 15 portions, how much venison leg cut into 5 ounce servings would be needed?
   a. 4.69 pounds
   b. 5.25 pounds
   c. 5.33 ounces
   d. 3.46 pounds

19D. Chapter Review

For each statement below, circle True or False to indicate the correct answer. If an answer is false, then explain why.

1. Most farmed deer is not slaughtered or processed in the slaughterhouse.
   True   False

2. Wild antelope, venison and rabbit are not subject to inspection under federal law.
   True   False

3. A mature boar (3-4 years old) has a better flavor than a baby boar.
   True   False

4. Wild game birds can be purchased by request from most butchers.
   True   False

5. Wild boar is closely related to the domestic pig.
   True   False

6. Game is higher in fat and vitamins than most other meats.
   True   False

7. Venison is very moist due to the marbling through the tissue.
   True   False

8. Large game animals are usually sold in primal portions.
   True   False

9. Furred game meat has a finer grain than other meats.
   True   False

10. The aroma, texture and flavor of game is affected by the lifestyle of the animal.
    True   False

11. Antelope, deer and rabbit are the game most widely available to food service operations.
    True   False

12. The flesh of game is generally moist and tender.
    True   False

13. During the hanging process carbohydrates convert into lactic acid, which tenderizes the flesh.

    True   False

14. Commercially raised game should *always* be marinated.

    True   False

15. Only a few species of furred game are widely available to food service operations.

    True   False

16. Beefalo steaks should always be cooked rare or medium rare.

    True   False

17. Farm raised pheasant usually comes dressed and seldom weighs more than 2.25 pounds.

    True   False

19E. Putting It All Together

Provide a short response for each of the questions below. These questions are designed to help you connect the bigger concepts presented in this chapter and/or text.

1. This chapter provides some information on Bison, saying that its meat is juicy and flavorful and may be prepared in the same manner as lean beef. From the knowledge you gained from Chapter Fourteen, suggest some ways that you might prepare Bison.

2. Think about all the four-legged animals that the text *On Cooking* has reviewed, ranging from beef, pork, lamb, and even game. Although names of cuts may vary among the different animals, what area of the animals' body tends to be most tender?

20

FISH AND SHELLFISH

> ## TEST YOUR KNOWLEDGE

The practice sets provided below have been designed to test your comprehension of the information found in this chapter. It is recommended that you read this chapter completely before attempting these questions.

20A. Terminology

Fill in the blank spaces with the correct definition.

1. En papillotte _____

2. Cephalopods _____

3. Mollusks _____

4. Anadromous _____

5. Aquafarmed _____

6. Bivalves _____

7. Pan-dressed _____

8. Univalves _____

9. Round fish _____

10. Whole or round _____

11. Crustaceans _____

12. Butterflied _____

13. Steak _____

14. Submersion poaching _____

15. Drawn _____

16. Cuisson _____

17. Fillet _____

18. Shallow poaching _____

19. Tranche _____

20. Flat fish _____

21. Wheel or center cut _____

20B. Multiple Choice

For each request below, choose the one correct response.

1. To maintain optimum freshness, fish and shellfish should be stored at what temperature?
    a.   40° F
    b.   30°-34° F
    c.   40°-45° F
    d.   38°-40° F

2. Fish are graded:
    a.   USDA Prime, Choice, Select or Utility
    b.   Type 1, type 2, type 3
    c.   Premium, Commercial Grade, Cutter/Canner
    d.   USDC A, B or C

3. Clams, mussels and oysters should be stored:
    a.   36° F
    b.   On ice and in refrigeration
    c.   In boxes or net bags
    d.   At 20% humidity

4.   Univalves and bivalves are both examples of:
    a.   Mollusks
    b.   Cephalopods
    c.   Clams
    d.   Crustaceans

5. The "universal" meaning of *prawn* refers to a:
    a.   Shrimp sautéed in garlic and butter
    b.   All shrimp, fresh water or marine variety
    c.   Shrimp from the Gulf of Mexico
    d.   Fresh water variety of shrimp only

6.   In terms of the market forms of fish, *dressed* refers to:
    a.   Viscera is removed
    b.   Viscera, gills, fins and scales are removed
    c.   As caught, intact
    d.   Viscera, fins and gills removed, scaled and tail trimmed

7. The most important commercial variety of salmon is:

   a. Atlantic

   b. Pacific

   c. Chinook

   d. King

8. Which type of sole cannot be caught off the coastline of the United States?

   a. Lemon

   b. English

   c. Petrale

   d. Dover

9. Mackerel, wahoo, herring, sardines and salmon have similar characteristics in that:

   a. The color of their flesh is the same

   b. They all migrate

   c. Their flesh is moderately-highly oily

   d. Their geographic availability is the same

10. All clams are examples of:

   a. Cephalopods

   b. Crustaceans

   c. Univalves

   d. Bivalves

11. The best selling fish in America is:

   a. Atlantic salmon

   b. Lemon sole

   c. Cod

   d. Ahi tuna

12. When cooking fish fillets with the skin on, what can be done to prevent the fillet from curling?

   a. Cook fillet at a high temperature, short time

   b. Cook fillet at a low temperature, longer time

   c. Score the skin of the fish before cooking

   d. Flatten the fillet by weighing down with a semi-heavy object during cooking

13. Which is the fastest method of freezing fish, therefore increasing the likelihood of the freshest thawed product?

   a. Glazed

   b. Frozen

   c. Fresh-frozen

   d. Flash-frozen

14. Assuming that fresh sides of salmon naturally have a shelf life of seven days, what will happen to their shelf life if stored in the refrigerator at 40°F/4°C? The shelf life will be decreased by:

    a. One-quarter

    b. One-third

    c. One-half

    d. Two-thirds

20C. Market Forms of Fish

Identify the market forms indicated in the following diagram and write their names in the spaces provided below.

1. _____

2. _____

3. _____

4. _____

5. _____

6. _____

7. _____

1.

2.

3.

4.

5.

6.

7.

20D. Short Answer

Provide a short response that correctly answers the following requests.

1. Give two (2) reasons why fish fillets and steaks are the best market forms to bake.

    a. _____

    b. _____

2. List the four (4) guidelines for determining the doneness of fish and shellfish.

    a. _____

    b. _____

    c. _____

    d. _____

3. List four (4) cooking methods that would be appropriate for preparing a tranche of salmon.

   a. _____

   b. _____

   c. _____

   d. _____

4. List two (2) oily and two (2) lean fish that grill well.

   Oily   **Lean**

   a. _____   a. _____

   b. _____   b._____

5. Name four (4) types of shellfish that are good for sautéing.

   a. _____   c. _____

   b. _____   d. _____

6. List three (3) dishes that exemplify why shellfish are good for baking.

   a. _____

   b. _____

   c. _____

7. Explain two (2) reasons why combination cooking methods are not traditionally used to prepare fish and shellfish.

   a. _____

   b. _____

8. List four (4) of the seven (7) quality points used to determine the freshness of fish.

   a. _____   c. _____

   b. _____   d. _____

20E. Chapter Review

For each statement below circle either True or False to indicate the correct answer. If an answer is false, then explain why.

1. All fish are eligible for grading.

   True   False

2. Fish and shellfish inspections are mandatory.

   True   False

3. A lobster is an example of a crustacean.

   True   False

4. Fatty fish are especially good for baking.

   True   False

5. The only difference between Maine lobsters and spiny lobsters is the geographic location where they're caught.

   True   False

6. Atlantic hard-shell clams are also known as geoducks.

   True   False

7. Fillet of halibut is a good fish to pan-fry.

   True   False

8. Cooking fish or shellfish en papillote is an example of baking.

   True   False

9. Shellfish has as much cholesterol as lamb.

   True   False

10. Salmon gets its pink-red flesh color from the crustaceans it eats.

    True   False

11. Halibut and sole are examples of flat fish.

    True   False

12. The best market form to purchase monkfish in is steaks.

    True   False

13. Surimi has equal nutritional value to the real fish and shellfish it replaces.

    True   False

14. Crustaceans are shellfish.

    True   False

15. The FDA permits the practice of marketing many types of flounder as sole.

    True   False

16. When preparing a Tilapia fillet to shallow poach, sprinkle it well with seasonings before submerging it in the cooking medium.

    True   False

17. Cod, sea bream, trout, scallops and shrimp are all popular items to sauté.

    True   False

18. Pollock is also known as Boston bluefish and has pink flesh when raw that turns white when cooked.

    True   False

19. To improve flavor, all fish should be brushed with butter or oil before broiling or grilling.

    True   False

20. Fish should be cooked to an internal temperature of 145°F for 15 seconds.

    True   False

20F. Putting It All Together

Provide a short response for each of the questions below. These questions are designed to help you connect the bigger concepts presented in this chapter and/or text.

1. Aside from the fact that most fish and shellfish are low in calories, fat and sodium and are high in protein and vitamins A, B and D, omega-3 fatty acids as well as many minerals, what else contributes to their healthfulness?

2. What are some things a chef has to consider when determining what market forms of fish/shellfish to purchase?

21

EGGS AND BREAKFAST

➤ TEST YOUR KNOWLEDGE

The practice sets provided below have been designed to test your comprehension of the information found in this chapter. It is recommended that you read this chapter completely before attempting these questions.

21A. Terminology

Fill in the blank spaces with the correct definition.

1. Soft boiled _____

2. Chalazae chords _____

3. Shell _____

4. Pasteurization _____

5. Sunny-side up _____

6. Yolk _____

7. Hard boiled _____

8. Egg white _____

9. Over easy _____

10. Basted eggs _____

11. Over hard _____

12. Over medium _____

21B. Multiple Choice

For each request below, choose the one correct response.

1. At what temperature does an egg yolk solidify (coagulate) when cooking?
   a. 135°F to 143° F
   b. 120°F to 132°F
   c. 160°F to 171°F
   d. 149°F to 158°F

2. What would be a good use for grade B eggs?
   a. As a compound in facial creams and other cosmetic products
   b. For baking, scrambling or the production of bulk egg products
   c. Grade B eggs are not recommended for use in food service operations
   d. For frying, poaching or cooking in the shell

3. What should a chef do to help the egg whites cling together when poaching an egg?
   a. Add salt to the water.
   b. Add a small amount of white vinegar or other acid to the water.
   c. Add the egg to the cooking liquid before it has a chance to simmer.
   d. Use only Grade A eggs.

4. In-shell cooking of eggs uses what method of cookery?
   a. Frying
   b. Steaming
   c. Boiling
   d. Simmering

5. Which *four* of the following are criteria for grading eggs?
   a. Certification of farmer
   b. Albumen
   c. Spread
   d. Shell
   e. Breed of bird
   f. Yolk

6. What is the maximum amount of time an egg dish can be left at room temperature (including preparation and service time) before it becomes potentially hazardous to consume?
   a. 20 minutes
   b. 40 minutes
   c. 1 hour
   d. 2 hours or more

7. Shirred eggs and quiche are both prepared by using a dry-heat cooking method. Which method do they have in common?

    a. Baking

    b. Roasting

    c. Frying

    d. Sautéing

8. Which is *false* regarding brunches? They:

    a. Are composed of a combination of breakfast and lunch items

    b. Are an eat-on-the-go meal experience

    c. May be accompanied by alcoholic beverages

    d. Are a social, leisurely meal

9. Which one of the following statements is *false* about the definition of a frittata? Frittatas are:

    a. Omelets containing a generous amount of ingredients that are folded in half when served

    b. Of Spanish-Italian origin

    c. Started on the stovetop, then placed in the oven, or under a salamander or broiler to finish cooking

    d. Different sizes, depending on the size pan used to prepare them

10. Cartons of fresh, uncooked (refrigerated) eggs in the shell are safe to use in cooking:

    a. Until the expiration date stamped on the package

    b. Until three weeks beyond the packing date stamped on the package

    c. Until four to five weeks beyond the packing date stamped on the package

    d. Up to three months after they are laid by the chicken

11. Egg yolks do not contain as much cholesterol as once feared. According to the American Heart Association, how many eggs per week can be consumed and still maintain a balanced diet? Up to:

    a. Two

    b. Four

    c. Six

    d. Ten

12. The internal temperature of a cheese blintze served on a brunch buffet should be:

    a. 145°F

    b. 155°F

    c. 165°F

    d. 175°F

21C. Egg Identification

a. _____

b. _____

c. _____

d. _____

Germinal spot

a.

c.

Air cell

d.

b.

21D. Coffees and Teas

Fill in the blanks provided with the response that correctly completes the statement.

1. Coffee can be judged on four characteristics:

   a. _____ Refers to the feeling of heaviness or thickness that coffee provides on the palate

   b. _____ Will often indicate the taste of coffee

   c. _____ Refers to the tartness of the coffee, lending a snap, life, or thinness

   d _____ The most ambiguous as well as the most important characteristic, having to do with taste

2. _____ is a commercial coffee bean from which the finest coffees are produced.

3. _____ is a bean which does not produce as flavorful a coffee, but is becoming more significant commercially since the trees are heartier and more fertile than their predecessors.

4. The best results for brewing a good cup of coffee are nearly always achieved by using _____ level tablespoons of ground coffee per 3/4 measuring cup (_____ ounces) of water.

5. A cup of _____ is often either the very first or the very last item consumed by a customer

6. Whether iced or hot, _____, is often consumed throughout the meal.

7. _____, _____ and _____ are the three basic types of tea.

21E. Chapter Review

For each statement below circle either True or False to indicate the correct answer. If an answer is false, then explain why.

1. Egg whites solidify (coagulate) when cooked at temperatures between 144° F and 149° F.
   True   False

2. Shell color has an effect on the grade of the egg, but not on flavor or nutrition.
   True   False

3. A French-style omelet is prepared by filling the egg mixture with a warm, savory mixture of ingredients while the eggs cook in the pan.
   True   False

4. The temperature of the cooking surface on which pancakes are made should be 225°F.
   True   False

5. To ensure maximum volume when preparing whipped egg whites, the whites should be thoroughly chilled prior to whipping.
   True   False

6. Eggs should be stored at temperatures below 30° F and at a relative humidity of 70 to 80 percent.
   True   False

7. Egg substitutes can replace whole eggs in all cooking applications.
   True   False

8. Egg whites contain cholesterol.
   True   False

9. Pancakes are ready to be flipped when the first side is golden brown or when bubbles appear on the pancakes' surface.
   True   False

10. Eggs used for pan-frying should be a high grade and very fresh since the yolk holds its shape better and the white spreads less.
    True   False

11. Waffles and pancakes are examples of quick breads.
    True   False

12. Eggs can absorb strong aromas while in storage.
    True   False

13. A custard is made by poaching a mixture of eggs, cream or milk and seasonings.
    True   False

14. Meat is seldom the main part of the meal at breakfast, but rather an accompaniment.
    True   False

15. If properly refrigerated, a carton of fresh, uncooked eggs will keep for at least 4 to 5 weeks beyond the pack date found on the carton.
    True   False

16. Waffle irons should be washed after each use to maintain a sanitary cooking surface.
    True   False

17. Eggs are considered a potentially hazardous food product.

    True   False

18. When making an egg white omelet, one should use the same amount of heat and cooking temperature as used when making a traditional omelet.

    True   False

19. Crepes are the same as pancakes.

    True   False

20. When preparing mise en place for breakfast service, a chef must be sure that cracked eggs and pancake, waffle, or crepe batters are stored in small batches on ice on the service line in order to maintain internal temperatures of 40°F or lower before cooking.

    True   False

21. Green tea is yellow-green in color and partially fermented to release its characteristics.

    True   False

22. Caffe latte is made by mixing 1/4 espresso with 3/4 steamed milk without foam.

    True   False

23. Whole coffee beans will stay fresh for a few weeks at room temperature whereas ground coffee will only stay fresh three or four days.

    True   False

24. Decoction is the oldest method of making coffee but is now used only when preparing extremely strong Turkish coffee.

    True   False

25. Bottled water should be poured over ice when served.

    True   False

26. Serving flavored coffees is a relatively new practice started by the Americans.

    True   False

27. Tisanes have a long history in this country, dating back to the American Revolution, and contain no real tea but rather a mixture that may contain herbs, dried flowers, seeds or roots.

    True   False

28. The tea bag was invented quite by accident when an American merchant sent samples of his teas sewed into small bags made of light material to customers. The customers, without instruction, placed the bags in a cup and poured hot water over them, creating an instant cup of tea that required no straining or mixing.

    True   False

21F. Putting It All Together

Provide a short response for each of the questions below. These questions are designed to help you connect the bigger concepts presented in this chapter and/or text.

1. How careful should one be to not consume too many eggs in one week due to the level of cholesterol they contain?

2. What food safety precautions should a chef take into consideration when handling eggs for breakfast service?

22

VEGETABLES

> ## TEST YOUR KNOWLEDGE

The practice sets provided below have been designed to test your comprehension of the information found in this chapter. It is recommended that you read this chapter completely before attempting these questions.

22A. Terminology

Fill in the blank spaces with the correct definition.

1. Parboiling_____

2. Vegetable _____

3. Beurre noisette _____

4. Cellulose _____

5. Blanching_____

6. Refreshing or shocking _____

7. Vegetarian _____

8. Fruitarian _____

9. Demi-vegetarian_____

10. Lacto-ovovegetarian or ovo-lactovegetarian _____

11. Lactovegetarian _____

12. Vegan _____

13. Tofu _____

14. Seitan _____

15. Tempeh _____

16. Textured Soy Protein _____

22B. Multiple Choice

For each request below, choose the one correct response.

1. A braised vegetable dish differs from a stewed vegetable dish in that it:

    a. Contains an acid product

    b. Is usually prepared with only one vegetable

    c. Has a longer cooking time

    d. Is served with a reduction of the cooking liquid

2. Grades for all vegetables include:

    a. U.S. No. 1, U.S. No. 2, U.S. No. 3

    b. U.S. Grade A, U.S. Grade B, U.S. Grade C

    c. U.S. Extra Fancy, U.S. Fancy, U.S. Extra No. 1, U.S. No. 1

    d. USDA Recommended, USDA Approved

3. Vegetables are considered savory because:

    a. They are an herbaceous plant that can be partially or wholly eaten

    b. They have less sugar than fruit

    c. They have little or no woody tissue

    d. They are usually eaten cooked, not raw

4. Which of the following is *false* about sautéing vegetables?

    a. The finished product should be firm to the bite, brightly colored and show little moisture loss.

    b. All preparation of ingredients should be done in advance because the cooking process proceeds rapidly.

    c. Seasonings should be added to the pan when warming the oil or fat so their flavors have time to develop during this quick cooking process.

    d. A wide variety of vegetables can be sautéed.

5. Which one of the following is a *disadvantage* when grilling vegetables?

    a. The high heat of the cooking process kills much of the nutritional content.

    b. The types of vegetable to be grilled must be carefully selected based on their cooking times.

    c. Vegetables often need to be brushed or marinated with a little oil or fat before cooking.

    d. Smaller vegetables should be skewered to make handling easier.

6. Which of the following vegetables is *not* suitable for roasting or baking?

    a. Eggplant

    b. Potatoes

    c. Spinach

    d. Peppers

7. Fresh vegetables are sold by:

    a. Cases or flats

    b. Degree of processing

    c. Weight or count

    d. Specifications

8. To preserve nutrients, color and texture:

    a. Cut the vegetables into uniform shapes before cooking

    b. Cook the vegetables whole, then peel and cut

    c. Add acid to the cooking liquid

    d. Cook the vegetables as little as possible

9. When pan-steaming vegetables:

    a. Overcooking is less likely to happen

    b. Cover the cooking apparatus to retain heat

    c. More nutrients are lost than in other techniques

    d. Choose only vegetables with a firm texture

10. Which of the following is true about microwave cooking?

    a. It substitutes well for all cooking techniques except broiling and grilling.

    b. It is best used as a substitute for traditional steaming.

    c. It is dangerous to use in large-scale food service operations.

    d. Its cooking process, agitating water molecules within food, depletes nutrients.

11. Fresh soy beans are:

    a. Purple

    b. Expensive

    c. Categorized as a fruit

    d. A potentially hazardous food product

12. Which is *false* regarding organic foods?

    a. The USDA regulates production and labeling.

    b. They may contain pesticides, fungicides, and/or herbicides.

    c. Foods labeled "100% organic" must only contain organic ingredients.

    d. They contain few if any intentional additives and are free of incidental additives.

22C. Product Identification

Match each vegetable in List A with the appropriate letter in List B. Each choice in List B can only be used once.

List A

_____ 1. Artichokes

_____ 2. Swiss chard

_____ 3. Okra

_____ 4. Bok choy

_____ 5. Pumpkin

_____ 6. Tomatillos

_____ 7. Leeks

_____ 8. Truffles

_____ 9. Corn

_____ 10. Cucumbers

_____ 11. Carrots

_____ 12. Hot peppers

_____ 13. Bean curd

_____ 14. Olive

List B

a. A winter squash variety especially popular in October.

b. A type of beet used only for its greens.

c. A firm, orange taproot that is eaten raw or cooked.

d. Tubers that grow near oak or beech tree roots.

e. A member of the *capsicum* family commonly used in Asian, Indian, Mexican, and Latin American cuisines.

f. From Arab and African cuisines, a pod often used for thickening.

g. A sweet, onion-flavored vegetable with flat, wide leaves.

h. A white-stemmed variety of southern Chinese cabbage.

i. Immature flowers of a thistle plant often canned or marinated.

j. Husk tomatoes with a crisp, tart flesh.

k. The immature stalks of bulb onions.

l. A squash that comes in two varieties: pickled and sliced.

m. A plant seed that is really a grain or type of grass; grows on a cob.

n. A cheese-like soybean product with high nutritional value, low cost and high flavor adaptability.

o. The fruit of a tree native to the Mediterranean area that is inedibly bitter and must be washed, soaked and cured or pickled before eating.

22D. Chapter Review

For each statement below circle either True or False to indicate the correct answer. If an answer is false, then explain why.

1. Although frozen vegetables are often colorful, their texture may be softer than fresh vegetables.

    True   False

2. Puréed vegetables are usually prepared by first sautéing, steaming or boiling.

    True   False

3. Winter squash is commonly braised or stewed due to its dense texture.

    True   False

4. Food is irradiated by exposing it to gamma rays to sterilize it, slow ripening or prevent sprouting.

    True   False

5. Eggplants, peppers and tomatoes are considered "fruit-vegetables."

    True   False

6. The larger the size of the chile pepper, generally the milder its flavor.

    True   False

7. Examples of legumes are dried beans and peas.

    True   False

8. The grading of vegetables is not required by the USDA.

    True   False

9. Potatoes, onions, shallots and garlic are best stored between 34° and 40° F.

    True   False

10. An acid added to the cooking liquid causes a vegetable to resist softening and therefore require a longer cooking time.

    True   False

11. Flavenoids are found mainly in beets, cauliflower and winter squash.

    True   False

12. Timing a vegetable as it cooks is the best way to determine doneness.

    True   False

13. The ripening of vegetables proceeds more rapidly in the presence of carbon dioxide gas.

    True   False

14. If the *only* goal is to help vegetables retain color when cooked, then an alkali is a good ingredient to add to the cooking liquid.

    True   False

15. The FDA classifies food irradiation as a preservative.

    True   False

16. Slicing eggplant and sprinkling it with salt is necessary when preparing to roast the vegetable.

    True   False

17. A Chipotle Chile Pepper is made by drying a Jalapeño.

    True   False

18. The excess liquid used to can vegetables is what causes the contents of the can to lose nutrients and the texture to soften.

    True   False

19. In South America and cooking of the southern United States, collard greens are most commonly prepared by slowly simmering until very tender.

    True   False

20. Cooking destroys much of the sweetness and special flavor of Walla-Walla, Vidalia and Maui onions.

    True   False

21. Seven 3-ounce servings of Braised Celery with Basil (recipe 22.8) would require approximately 12 fresh basil leaves cut Chiffonade.

    True   False

22. Tofu is a potentially hazardous food product.

    True   False

23. Vegetables are valuable in our daily diets due to the carbohydrates, fiber, vitamins and minerals they provide.

    True   False

24. Vegetables are braised or stewed for the primary purpose of tenderizing.

    True   False

25. When soaking dried beans, allow 2 cups of cold, unseasoned liquid for each cup of dried beans.

    True   False

22E. Putting It All Together

Provide a short response for each of the questions below. These questions are designed to help you connect the bigger concepts presented in this chapter and/or text.

1. Based on your knowledge of nutrition gleaned from Chapter Three and your new study of this chapter on vegetables, create some healthy rules for incorporating a variety of vegetables into your daily diet.

2. This chapter talks about applying various cooking methods to vegetables and discusses the colors of the vegetables and what might happen to those colors when they're cooked under certain conditions. Relate the information on color presented in this chapter with the information presented on color in Chapter Seven, Flavors and Flavorings. What are the similarities and differences of the messages?

POTATOES, GRAINS AND PASTA

➢ TEST YOUR KNOWLEDGE

The practice sets provided below have been designed to test your comprehension of the information found in this chapter. It is recommended that you read this chapter completely before attempting these questions.

23A. Terminology

Fill in the blank spaces with the correct definition.

1. Extruded _____

2. Sfoglia _____

3. Tossing method _____

4. Converted rice _____

5. New potatoes _____

6. Dumpling _____

7. Hulling _____

8. Medium-grain rice _____

9. Endosperm _____

10. Filled dumpling _____

11. Berry _____

12. Durum wheat _____

13. Mealy potatoes _____

14. Plain/drop dumpling _____

15. Germ _____

16. Still-frying method_____

17. Groat _____

18. Waxy potatoes _____

19. Hull _____

20. Short-grain rice _____

21. Instant/quick cooking rice _____

22. Cracking_____

23. Brown rice _____

24. Masa harina _____

25. Pearling _____

26. White rice_____

27. Long-grain rice_____

28. Grinding_____

29. Bran _____

23B. Short Answer

Provide a short response that correctly answers each of the following requests.

1. Why is it so important to use ample water when cooking pasta?

   _____

2. Why shouldn't a baked potato be cooked by wrapping in foil or microwaving?

   a. _____

   b. _____

3. Name three (3) dishes that are traditionally made with short-grained rice.

   a. _____

   b. _____

   c. _____

4. The finest commercial pastas are made with pure semolina flour. Why?

   _____

5. Duchesse potatoes are considered the mother to many classical potato dishes. List four (4)
   different classical dishes prepared from Duchesse, briefly describing the ingredients.

   a. _____

   b. _____

   c. _____

   d. _____

6. Identify the three (3) main shapes of Italian pasta.

   a. _____   c. _____

   b. _____

7. What are the three (3) basic cooking methods for cooking grains?

   a. _____   c. _____

   b. _____

8. Give three (3) reasons for soaking most dried Asian noodles in hot water before cooking:

   a. _____

   b. _____

   c. _____

23C. Multiple Choice

For each request below, choose the one correct response.

1. Which of the following classical potato preparations include Duchesse as part of the ingredients?

   a. Anna

   b. Berny

   c. Rosti

   d. Boulangère

2. Which of the following grains *cannot* be used to make risotto?

   a. Barley

   b. Oats

   c. Buckwheat

   d. Arborio rice

3. American-grown rice *does not* need to be rinsed before cooking because:

   a. All of the starch will be washed away

   b. The rice will become soggy before cooking

   c. Rinsing will result in a sticky rice

   d. Such rice is generally clean and free of insects

4. When boiling pasta, "ample water" is defined by measurements:

   a. 1 quart of water to 1 pound of pasta

   b. 2 quarts of water to 1 pound of pasta

   c. 15:1 ratio of water to pasta

   d. 1 gallon of water to 1 pound of pasta

5. Which of the following is *false* about converted rice? It:
    a. Tastes the same as regular milled white rice
    b. Retains more nutrients than regular milled white rice
    c. Has been pearled in order to remove the surface starch
    d. Cooks more slowly than regular milled white rice

6. Why is long grain rice more versatile and popular than other types of rice? Long grain rice:
    a. Has a higher nutritional content than short or medium grained rice
    b. Remains firm and separate when cooked properly
    c. Is far more affordable for a larger variety of foodservice operations
    d. Is easier and faster to cook than other rice

7. Which type of potato would be good for making Potatoes Berny?
    a. Mealy potatoes
    b. Waxy potatoes
    c. New potatoes
    d. Sweet potatoes

8. Which statement is *false* about the nutritional content of grains? They:
    a. Contain all of the essential amino acids
    b. Are high in fat
    c. Are a good source of dietary fiber
    d. Are a good source for vitamins and minerals

9. Which of the following flours is used to make Asian noodles?
    a. Potato
    b. Bean
    c. Corn
    d. Oat

10. Duchesse potatoes that are shaped, breaded and fried result in:
    a. Croquettes
    b. Dauphine
    c. Marquis
    d. Lorette

11. What is the difference between cooking fresh pasta and dry, factory-produced pasta?
    a. Fresh pasta takes significantly less time to cook
    b. Dried pasta should be cooked to order
    c. Dried pasta takes significantly less time to cook
    d. Fresh pasta contains a much different list of ingredients

12. How hot should the fat be for blanching potatoes?
    a. 250 to 300°F
    b. 275 to 325°F
    c. 300 to 350°F
    d. 350 to 375°F

13. Which grain contains all 9 essential amino acids and therefore is considered to be a complete protein?
    a. Brown rice
    b. Barley
    c. Millet
    d. Quinoa

23D. Chapter Review

For each statement below circle either True or False to indicate the correct answer. If an answer is false, then explain why.

1. Grains cooked by the risotto or pilaf method are first coated with hot fat.
    True   False

2. Medium-grain rice is best when freshly made and piping hot.
    True   False

3. The only grain eaten fresh as a vegetable is corn.
    True   False

4. Making a dough with semolina flour makes it softer, more supple and easier to work with.
    True   False

5. Asian noodle dough can be used to make dumplings.
    True   False

6. "Yam" is an industry term for sweet potato.
    True   False

7. Potatoes should be stored between 40° and 50° F.
    True   False

8. The best applications for mealy potatoes are sautéing and pan-frying.
    True   False

9. A ravioli is a dumpling.
    True   False

10. Fresh pasta is best when cooked to order.
    True   False

11. Top quality Russet potatoes are recommended for deep-frying.
    True   False

12. Three basic cooking methods are used to prepare grains: simmering, risotto and pilaf.
    True   False

13. The standard ratio for cooking rice is 1 part liquid to 1 part rice.

    True   False

14. Pasta is widely used in the cuisines of Asia, North America, and Europe.

    True   False

15. Generally, cracked wheat and bulgur can be substituted for one another in recipes.

    True   False

16. Buckwheat is neither wheat nor grain.

    True   False

17. The high pH and high protein content of rice make it a potentially hazardous food product.

    True   False

18. One cup of dry quinoa will yield approximately one cup of cooked.

    True   False

19. Dumplings are always stuffed.

    True   False

20. Al dente means firm but tender.

    True   False

21. Potato and grain dishes are potentially hazardous foods and should be heated to 145ºF and then held for service at 135ºF.

    True   False

23E. Putting It All Together

Provide a short response for each of the questions below. These questions are designed to help you connect the bigger concepts presented in this chapter and/or text.

1. This chapter and the Chapter 22 on vegetables, keep emphasizing to beware that vegetables and potatoes, pastas and grains are potentially hazardous foods. Explain again the characteristics of potentially hazardous foods and why there seems to be such a heavy warning in these two particular chapters.

2. Building off your answer to the previous question, explain why wrapping a baked potato in aluminum foil during the baking and holding process during service (even if the proper holding temperature is maintained) can be so dangerous.

24

VEGETARIAN COOKING

> ## TEST YOUR KNOWLEDGE

The practice sets provided below have been designed to test your comprehension of the information found in this chapter. It is recommended that you read this chapter completely before attempting these questions.

24A. Terminology

Fill in the blank spaces with the correct definition.

1. Pythagoreans _____

2. Strict or pure vegetarian _____

3. Living foodist _____

4. Macrobioticist _____

5. Silken tofu _____

6. Miso _____

7. Analogous foods _____

8. Cotton tofu _____

24B. Vegetarianism as Determined By Religion

Below is a list of numbered beliefs related to vegetarianism and major world religions. Under each numbered fact list the letter of the religion(s) that prescribe to the belief. More than one religion may be matched with each belief.

A. Hinduism   B. Buddhism   C. Jainism   D. Judaism & Christianity

_____ 1. Some follow a religious and ethical standard of ahimsa

_____ 2. Practices can vary by geographic location grouping such as China and Vietnam; Japan and Korea; Sri Lanka and Southeast Asia

_____ 3. Majority live in India

_____ 4. Approximately 800 million observe this faith worldwide

_____ 5. Eating meat is common

_____ 6. Based on Vedas, sacred texts approximately 4000 years old

_____ 7. Vegetarianism is common, but not followed by all

_____ 8. Some follow ancient texts that believe Buddha ate meat because he didn't want to offend those who brought charitable offerings

_____ 9. Some prepare for the coming of the Messiah by practicing vegetarianism

_____10. Combines Hinduism and Buddhism

_____11. Try to practice as pure a form of ahimsa as possible, sweeping the ground before they walk on it and wearing gauze masks so as not to tread or breath on insects and harm them

_____12. Some believe they can eat meat

_____13. All follow the Dalai Lama, the spiritual leader worldwide

_____14. Most observers abstain from eating the sacred cow

_____15. Predominant religion in India

24C. Short Answer

Provide a short response that correctly answers the following requests.

1. Name the one vitamin and one mineral either not found in vegetables or the type found in the plant food is not readily absorbed by the body, therefore requiring a supplement.

    Vitamin: _____

    Mineral: _____

2. List four (4) reasons why one might choose a vegetarian lifestyle.

    a. _____

    b. _____

    c. _____

    d. _____

3. Frances Moore Lappé's 1971 publication supported vegetarianism from what perspective?

    _____

    _____

4. Dieticians recommend that vegans include linolenic acids in their diets so as to compensate for fatty acids that may be deficient. What four ingredients offer a good supply of such acids?

    a. _____   c. _____

    b. _____   d. _____

5. Australian ethics professor, Peter Singer, wrote a book in 1975 that helped spark the United States-based PETA (People for the Ethical Treatment of Animals). What two things does this organization stand for?

a. _____

b. _____

6. What makes vegetarian diets so hard for a chef to understand when trying to appeal to guests' desires?

_____

_____

_____

7. Vegans can meet the calcium requirements in their diets by consuming what four things?

a. _____   c. _____

b. _____   d. _____

8. Explain the difference between white and red miso.

_____

_____

24D. Chapter Review

For each statement below circle either True or False to indicate the correct answer. If an answer is false, then explain why.

1. Compared to years past vegetarianism is gaining acceptability in the United States for more reasons than just religious or philosophical beliefs.

True   False

2. It is estimated that today nearly 25 million Americans forego some or all animal products in their diets.

True   False

3. Vegetarians who consume dairy and/or eggs generally have an easier time meeting their nutrient needs compared to a Fruitarian.

True   False

4. Tossed salads are a good source of quality protein.

True   False

5. Eggs are consumed by ovo vegetarians as well as vegans.

True   False

6. Grains can be consumed by people observing all types of vegetarian diets.

True   False

7. The same quantity of soy "milk" can be substituted for dairy milk in all recipes.

True   False

8. Peas, most nuts and seeds as well as oats are the only plant based foods that are equivalent to animal proteins.

    True   False

9. Once opened, soy milk that is sold in aseptic packaging (stored at room temperature) is good for one year as long as it is refrigerated.

    True   False

10. Nuts and seeds can be consumed by all types of vegetarians.

    True   False

11. Tempeh should be cooked prior to eating, which tempers its pronounced flavor.

    True   False

12. Of all the vegan protein ingredients, tofu has a texture similar to wheat.

24E. Putting It All Together

Provide a short response for each of the questions below. These questions are designed to help you connect the bigger concepts presented in this chapter and/or text.

1. Some see the bland flavor of tofu as being a negative, but based on your knowledge of Chapter Seven, speak to this characteristic of this soy product.

2. How can your knowledge of meat and meat cookery be used to understand ingredient substitutions for vegan dishes that use substitute proteins such as tofu, texturized soy protein, grain or bean puree?

25

SALADS AND SALAD DRESSINGS

> ## TEST YOUR KNOWLEDGE

The practice sets provided below have been designed to test your comprehension of the information found in this chapter. It is recommended that you read this chapter completely before attempting these questions.

25A. Terminology

Fill in the blank spaces with the correct definition.

1. Dressing _____

2. Bound salads _____

3. Mesclun _____

4. Base _____

5. Basic French dressing _____

6. Vegetable salads _____

7. Emulsified sauce _____

8. Green salads _____

9. Garnish_____

10. Composed salad_____

11. Body _____

12. Tossed salad_____

13. Fruit salad _____

14. Microgreens_____

25B. Multiple Choice

For each request below, choose the one correct response.

1. What is the best type of oil to use when making mayonnaise?
   a. Nut oils
   b. Vegetable oils
   c. Seed oils
   d. Olive oil

2. What are the *two* forms in which lettuce grow?
   a. Bunch and leaf
   b. Leaf and head
   c. Head and stalks
   d. Stalks and bunch

3. Lettuces and salad greens should be stored in protective containers at what temperature?
   a. 30° to 32° F
   b. 32° to 34° F
   c. 34° to 38° F
   d. 40° to 50° F

4. What type of an emulsion is a basic vinaigrette?
   a. Permanent
   b. Semi-permanent
   c. Temporary
   d. Semi-temporary

5. Approximately how much oil can one egg yolk emulsify?
   a. 2 ounces
   b. 4 ounces
   c. 1 cup
   d. 7 ounces

6. In a composed salad, the green would serve as the:
   a. Base
   b. Body
   c. Garnish
   d. Dressing

7. "Tomato and Asparagus Salad with Fresh Mozzarella" would be considered a:
   a. Fruit salad
   b. Ccomposed salad
   c. Vegetable salad
   d. Bound salad

8. Traditional potato salad is considered a:
   a. Green salad
   b. Composed salad
   c. Vegetable salad
   d. Bound salad

9. Due to the flavor characteristics of mâche, what *would not* be an appropriate green to toss with this in a salad?
   a. Boston lettuce
   b. Radiccio
   c. Bibb lettuce
   d. Iceberg lettuce

10. Radiccio, escarole, frisée, Belgian endive and dandelion are all examples of:
    a. Lettuce
    b. Sprouts
    c. Fresh herbs
    d. Bitter greens

11. From a food safety standpoint extra care should be taken when preparing salads because:
    a. Certain ingredients are expensive and should not be wasted
    b. Salads are usually foods that will not be cooked before service
    c. Salad ingredients have mild flavors that may easily become tainted by strong odors
    d. They wilt quickly and easily

12. If a temporary emulsion separates during service, the chef should:
    a. Re-whisk it immediately before use
    b. Whisk in one egg yolk for every 7 ounces of oil used in the recipe
    c. Whisk in 1/3 more oil
    d. Whisk in 2 tbsp Dijon mustard

25C. Short Answer

Provide a short response that correctly answers the following requests.

1. What are three (3) things that should be avoided when making a nutritionally balanced salad? The overuse of:

   a. _____

   b. _____

   c. _____

2. Give two (2) reasons why greens should be stored separately from tomatoes and apples.

   a. _____

   b. _____

3. List five (5) of the ingredients that may be included in mayonnaise-based dressings.

a. _____   d. _____

b. _____   e. _____

c. _____

4. Briefly describe the eight (8) basic steps to be followed when making mayonnaise.

a. _____

b. _____

c. _____

d. _____

e. _____

f. _____

g. _____

h. _____

5. List four (4) possible ingredients for a fruit salad dressing.

a. _____   c. _____

b. _____   d. _____

25D. Chapter Review

For each statement below circle either True or False to indicate the correct answer. If an answer is false, then explain why.

1. The balance of vinegar, oil, lecithin and whipping is crucial to achieve a proper emulsion.

   True   False

2. Chicory, Belgian endive, sorrel and spinach are all examples of salad greens that can be eaten raw or cooked.

   True   False

3. Romaine or Cos lettuce benefits from hand-tearing to break it into smaller pieces, while butterhead and baby lettuces can be cut with a knife.

   True   False

4. Tossed salads should be dressed at the last possible moment before service to prevent over-marination from the dressing.

   True   False

5. Generally softer-leaved varieties of lettuces like iceberg and red-leaf tend to perish more quickly in storage than crisper-leaved varieties.

   True   False

6. The best rule of thumb to follow when matching dressings to salad greens is, "the milder the flavor of the salad green, the milder the flavor of the dressing."

   True   False

7. All greens should be washed *after* they've been torn or cut.

   True   False

8. The standard ratio of oil to vinegar in a temporary emulsion is 2 parts to 1; however this ratio may vary when using strongly flavored oils, thus decreasing the proportion of oil to vinegar to 1 part oil to 1 part vinegar.

   True   False

9. Once washed, salad greens should be dried well to maintain a crisp texture and to ensure that oil-based dressings will adhere to the leaves.

   True   False

10. An advantage of using an emulsified vinaigrette dressing instead of a mayonnaise-based dressing is that it has the basic flavor of a vinaigrette without being as heavy as mayonnaise.

    True   False

11. Nasturtiums, miniature roses, and zinnias are all examples of edible flowers grown to enhance food presentations.

    True   False

12. Although salads are generally believed to be healthful foods, they can easily become laden with fat and calories based on the choice of dressings and toppings/garnishes.

    True   False

13. The conversion factor for producing Niçoise Salad (recipe 24.5) is 9.17 and therefore 48 portions will result.

    True   False

14. Raw egg yolks are safe to use when making homemade mayonnaise.

    True   False

15. A vinaigrette can serve as a dressing, a light sauce or a marinade.

    True   False

25E. Putting It All Together

Provide a short response for each of the questions below. These questions are designed to help you connect the bigger concepts presented in this chapter and/or text.

1. What obligation does a chef have to offer vegetarian and/or low fat salad options on the menu?

2. A common misconception about salads is that they are always a healthful, balanced way to eat. Look at the nutritional analysis for recipe 25.5, Salad Niçoise and discuss what you find.

26

FRUITS

> ## TEST YOUR KNOWLEDGE

The practice sets provided below have been designed to test your comprehension of the information found in this chapter. It is recommended that you read this chapter completely before attempting these questions.

26A. Terminology

Fill in the blank spaces with the correct definition.

1. Ripened _____

2. Acidulation _____

3. Preserve _____

4. Pectin _____

5. Marmalade _____

6. Papain_____

7. Gel _____

8. Jam _____

9. Jelly _____

10. Varieties_____

11. Hybrids_____

12. Respiration rate _____

13. Zest_____

26B. Short Answer

Provide a short response that correctly answers each of the following requests.

1. List the four (4) fruits that emit ethylene gas.

    a. _____   c. _____

    b. _____   d. _____

2. What is an indicator of cold damage to bananas?

    _____

3. Fruits are varied in their content of vitamins and minerals. Identify the fruits that are plentiful in the listed elements.

Vitamin C	Vitamin A	Potassium
1. _____	1. _____	1. _____
2. _____	2. _____	2. _____
3. _____ .	3. _____	3. _____

4. List four (4) uses for lower grades of fruit.

    a. _____   c. _____

    b. _____   d. _____

5. List five (5) methods of fruit preservation.

    a. _____   d. _____

    b. _____   e. _____

    c. _____

6. Name four (4) fruits that benefit from acidulation.

    a. _____   c. _____

    b. _____   d. _____

7. Name five (5) fruits that maintain their texture when sautéed.

    a. _____   d. _____

    b. _____   e. _____

    c. _____

26C. Fill in the Blanks

Fill in the blanks provided with the response that correctly answers the statement.

1. _____is the most common method of cooking pears.

2. In classical dishes, the term *à la Normande* refers to the use of _____.

3. Pumpkins, cucumbers and melons are all members of the _____ family.

4. _____ are the single largest fruit crop in the world.

5. When deep-frying fruits, the best results are achieved by first dipping the fruit slices in _____ before submerging in the hot fat.

26D. Product Identification

Match each fruit in List A with the appropriate letter in List B. Each choice in List B can only be used once.

List A

_____ 1. Tangerines

_____ 2. Grapes

_____ 3. Quince

_____ 4. Lemons

_____ 5. Sour cherries

_____ 6. Dates

_____ 7. Rhubarb

_____ 8. Plantains

_____ 9. Star Fruits

_____ 10. Pomegranates

List B

a. A vegetable that is prepared as a fruit, using lots of sugar to offset tart flavor

b. Also known as mandarins

c. Member of the gourd family

d. Too astringent to eat raw, but great when sprinkled with sugar

e. The single largest fruit crop in the world

f. Deep golden-yellow color and a full floral aroma when ripe

g. Appear to be dried, but are actually fresh fruits cultivated since ancient times

h. Unpleasant to eat raw, but great for flavoring savory foods and sweets

i. Light to dark red and so acidic they are rarely eaten uncooked

j. Concentrated juice is made into grenadine syrup

k. Larger but not as sweet as bananas, often cooked as a starchy vegetable

26E. Chapter Review

For each statement below circle either True or False to indicate the correct answer. If an answer is false, then explain why.

1. Carry-over cooking occurs with fruit.

   True   False

2. The two primary methods for juicing are pressure and blending.

   True   False

3. Sulfur dioxide is added to dried fruits to maintain their flavor during storage.

   True   False

4. Freezing is the best method of preserving the fresh appearance of fruit.

   True   False

5. The highest grade of fruit is U.S. No 1.

   True   False

6. Pineapples don't ripen after picking.

   True   False

7. Irradiation maintains fruit's flavor and texture while slowing the ripening process.

    True   False

8. Papayas are also known as carambola.

    True   False

9. Tropical fruit flavors complement rich or spicy meat, fish and poultry dishes.

    True   False

10. Papayas are ripe when a greater proportion of the skin is yellow rather than green.

    True   False

11. Red Delicious apples are good for making pies.

    True   False

12. Stone fruits such as mangoes are commonly dried, or made into liqueurs and brandies.

    True   False

13. Nutritionally speaking, fruits are low in protein and fat, high in fiber and a good source of energy.

    True   False

14. Meat tenderizers often contain enzymes similar to those found naturally in the seeds of kiwis, papayas and passion fruit.

    True   False

15. Fresh fruits are sold by weight or count.

    True   False

16. Although canning makes fruit's texture soft, it has little or no effect on vitamins A, B, C and D.

    True   False

17. Fruits laid in a pan and sprinkled with a strudel topping and then baked are called cobblers.

    True   False

18. Raspberries, blueberries and strawberries are often picked prior to fully ripening in order to extend shipping time.

    True   False

19. A fruit should be refrigerated in order to retard ripening once it has been harvested.

    True   False

20. Peaches, with the stone removed, are excellent as a "container" to be stuffed and baked.

    True   False

26F. Putting It All Together

Provide a short response for each of the questions below. These questions are designed to help you connect the bigger concepts presented in this chapter and/or text.

1. Citrus juices are often added to marinades and add great flavor but how can they help preserve food or at least slow bacterial growth?

2. In most cases what is the most nutritiously beneficial way of consuming fruits?

3. What is the big deal about heirloom varieties of fruit and will chefs eventually get back to using more readily available produce?

27

SANDWICHES

The practice sets provided below have been designed to test your comprehension of the information in this chapter. It is recommended that you read this chapter completely before attempting these questions.

27A. Terminology

Fill in the blank spaces with the correct definition.

1. Hot closed sandwiches _____

2. Hot open-faced sandwiches _____

3. Cold closed sandwiches _____

4. Cold open-faced sandwiches _____

27B. Short Answer

Provide a short response that correctly answers each of the following requests.

1. The cardinal rule of food handling is; keep hot foods _____ and keep cold food _____.

2. Sandwiches are especially prone to food-borne illness because of the use of high _____ foods.

3. The deadliest source of cross-contamination is _____ _____.

4. The three principal spreads are _____ , _____ and _____ _____.

5. Tuna salad is an example of a _____ salad.

6. Pizza is an example of a _____ , _____-_____ sandwich.

7. Name and describe the three (3) steps for sandwich preparation.

a._____

b. _____

c._____

27C. Multiple Choice

For each request question below, choose one correct response.

1. The purpose of a sandwich spread is to add:
   a. Flavor
   b. Color
   c. Texture
   d. Mouth feel

2. Which of the following is *not* a sandwich filling?
   a. Cheese
   b. Shellfish
   c. Vegetable purée
   d. Eggs

3. A Monte Cristo sandwich is an example of which of the following sandwiches?
   a. Wrap
   b. Hot open-faced
   c. Multidecker
   d. Deep-fried

4. Which of the following is an example of a hot open-faced sandwich?
   a. Pizza
   b. Quesadillas
   c. Tacos
   d. Tea sandwiches

5. All of the following are important to consider when choosing bread for a sandwich *except*:
   a. Dense breads are easier to slice
   b. Freshness and flavor
   c. Overly crusty breads are harder to eat
   d. Texture should withstand moisture from fillings and spreads

27D. Matching

Match the regional American sandwich name in List A with its description in List B. Choices in List B may only be used once.

List A

_____1. Hoagie

_____2. Grinder

_____3. Po' Boy

_____4. Hero

_____5. Muffuletta

List B

a. Sandwich soaked in gravy

b. New England sandwich that is a workout for the jaw

c. Name given to huge sandwiches by a food journalist

d. Sandwich made in a plancha

e. Sandwich from Philadelphia

f. Round sandwich whose bread is coated with a garlicky green-olive spread

27E. Chapter Review

For each statement below circle either True or False to indicate the correct answer. If an answer is false, then explain why.

1. Vegetable purées provide a barrier to prevent the bread from getting soggy.
    True   False

2. A club is an example of a multidecker sandwich.
    True   False

3. Chicken salad is an example of a spread.
    True   False

4. Tea sandwiches are a small version of a cold closed sandwich.
    True   False

5. Sandwich ingredients should be covered to prevent dehydration.
    True   False

6. Hamburgers are usually served closed so that the burger stays warm.
    True   False

7. An example of a cold open-faced sandwich is a bagel with lox and cream cheese.
    True   False

8. A gyro is made by placing a ground beef patty with grilled onion and cheese between 2 slices of buttered rye bread.
    True   False

9. Ingredients for sandwiches should always be stored at room temperature.
    True   False

10. A Reuben contains sliced roast beef, Swiss cheese, sauerkraut, and mustard or Thousand Island dressing on rye bread.
    True   False

27F. Putting It All Together

Provide a short response for each of the questions below. These questions are designed to help you connect the bigger concepts presented in this chapter and/or text.

1.   What modifications can a chef perform to make a sandwich more nutritionally sound?

2.   What should chefs packing sandwiches "to go" for guests be aware of regarding food safety?

28

CHARCUTERIE

> ## TEST YOUR KNOWLEDGE

The practice sets provided below have been designed to test your comprehension of the information found in this chapter. It is recommended that you read this chapter completely before attempting these questions.

28A. Terminology

Fill in the blank spaces with the correct definition.

1. Forcemeat_____

2. Dominant meat_____

3. Canadian bacon _____

4. Collagen casings _____

5. Pâté spice _____

6. Liver terrines _____

7. Smoker_____

8. Country-style forcemeats _____

9. Prosciutto _____

10. Cold smoking _____

11. Fat _____

12. Vegetable terrines _____

13. Panada _____

14. Mousse_____

15. Common bacon _____

16. Foie gras terrines _____

17. Hot smoking _____

18. Chopped chicken liver _____

19. Mousseline forcemeats _____

20. Curing salt _____

21. Dried or hard sausages _____

22. Boneless or formed hams _____

23. Westphalian ham _____

24. Fresh sausages _____

25. Basic forcemeats _____

26. Smoked and cooked sausages _____

27. Pâté _____

28. Country ham _____

29. Natural casings _____

30. Terrines _____

31. Pâtés en croûte _____

32. Fresh ham _____

33. Pancetta _____

34. Galantine _____

35. Brawns or aspic terrines _____

36. Rillette _____

37. Confit _____

38. Ballottine _____

28B. Short Answer

Provide a short response that correctly answers each of the following requests.

1. Name three (3) kinds of forcemeat that can be used to make a pâté en croûte.

    a. _____

    b. _____

    c. _____

2. If a forcemeat won't emulsify in a warm kitchen, what can be done?

    _____

3. Compare and contrast a galantine and a ballottine.

Galantine	**Ballottine**
a. _____	_____
b. _____	_____
c. _____	_____
d. _____	_____
e. _____	_____
f. _____	_____

4. What are three (3) steps one can take to ensure proper emulsification of a forcemeat?

a. _____

b. _____

c. _____

5. List five (5) reasons to use an aspic jelly.

a. _____

b. _____

c. _____

d. _____

e. _____

28C. Multiple Choice

For each request below, choose the one correct response.

1. When making a forcemeat, ingredient as well as equipment temperatures should be kept at what temperature throughout preparation?
    a. Below 41° F
    b. As cold as possible
    c. Room temperature, but not to exceed 60° F
    d. 43° to 45° F

2. Which of the following is *false* about salt curing? It:
    a. Inhibits bacterial growth
    b. Dehydrates the food
    c.   Is quick and easy
    d. Can take the place of cooking

3. Meat-based galantines, terrines and pâtés en croûte should be cooked to an internal temperature of:
    a. 145 ° F
    b. 125° F
    c. 150 ° F
    d. 160 ° F

4. When meats are cold smoked, what process is usually performed prior to the smoking?
   a. Salt curing and brining
   b. Trimming
   c. Barding
   d. Marinating

5. What gives ham, bacon and other smoked meats their pink color?
   a. Red dye #7 is added to the curing process
   b. The meat is cooked to a medium rare state of doneness
   c. Smoking, when done slowly, maintains the natural color of meats
   d. Nitrites are added to the cure

6. Which of the statements is *false* about a panada?
   a. Aids in emulsification
   b. Adds significant flavor
   c. Should not make up more than 20% of the forcemeat
   d. Is a binder

7. A chef is preparing a batch of pâté spice (recipe 27.1) and she's determined that she needs a total of 26.85 ounces. The conversion factor would be:
   a. 3.0
   b. 2.86
   c. 3.5
   d. 3.37

8. Which of the following would *not* be an appropriate garnish for a forcemeat?
   a. Whole, shelled and toasted pistachios
   b. Ground pork
   c. Brunoise carrot
   d. Batonette of tongue

9. A classically prepared aspic would include all of the following *except*:
   a. A strong flavored meat stock
   b. Fortification with gelatin
   c. Clarification
   d. Acid ingredient to balance flavor

10. Costing a recipe:
    A chef wants to produce 3.75 pounds of pâté dough (recipe 27.5). Based on the following information, what would be the ingredient cost?

    | A.P. Flour: | $.30 per pound | Shortening: | $.78 per pound |
    | Salt: | $.05 per ounce | Egg: | $1.00 per dozen |

    a. Flour: $.60; shortening: $.68; salt: $.03; egg: $.16
    b. Flour: $.30; shortening: $.34; salt: $.01; egg: $.08
    c. Flour: $.45; shortening: $.34; salt: $.02; egg: $.16
    d. Flour: $.75; shortening: $.88; salt: $.03; egg: $.20

28D. Matching

Match each of the terms in List A with the appropriate letter definition in List B.   Each choice in List B can only be used once.

List A

_____1. Mousse

_____2. Country-style forcemeat

_____3. Brawn

_____4. Forcemeat

_____5. Mousseline forcemeat

_____6. Pâté en croûte

_____7. Galantine

_____8. Quenelle

_____9. Terrine

List B

a. Pâté cooked in pastry dough

b. A cooked, light, airy, delicately flavored forcemeat

c. A poached dumpling of mousseline forcemeat

d. Meat, fish or poultry, bound, seasoned, with or without garnishes

e. A terrine made from highly simmered gelatinous cuts of meat, wine and flavoring

f. A purée of fully cooked meats, poultry, game, fish, shellfish or vegetables, lightened with cream and bound with aspic

g. A whole poultry item boned, stuffed and reshaped, poached and served cold

h. A deboned, stuffed poultry leg, poached or braised usually served hot

i. A hearty, highly seasoned, coarse textured forcemeat

j. A coarse forcemeat cooked in an earthenware mold

28E. Chapter Review

For each statement below circle either True or False to indicate the correct answer. If an answer is false, then explain why.

1. It is possible to make a vegetable mousseline forcemeat.
    True   False

2. The best type of mold to use to make a pâté en croûte is a metal loaf pan.
    True   False

3.   Sausages are forcemeat stuffed into casings.
    True   False

4. Béchamel sauces are used as a primary binding agent in most styles of forcemeats.
    True   False

5. Pork bellies are usually made into bacon.
    True   False

6. When marinating forcemeat ingredients before grinding, the trend today is to marinate them for longer periods to kill bacteria.
    True   False

7. Galantines are always served cold.
    True   False

8. Any type of forcemeat can be used to make a terrine.
   True   False

9. A fresh ham is made from the hog's shoulder.
   True   False

10. After testing a forcemeat's texture and finding it too firm, a little egg white should be added to fix the problem.
    True   False

11. A mousseline forcemeat can be served hot or cold.
    True   False

12. Only hams made in rural areas can be called country hams; all others must be called country-style hams.
    True   False

13. Chopped chicken liver has a longer shelf life than a rillette.
    True   False

14. Brining and pickling are the same procedure.
    True   False

15. The stock used to make chaud-froid sauce often determines the color and flavor of the sauce.
    True   False

16. Forcemeats, both raw and cooked, are considered potentially hazardous food products.
    True   False

17. A vegetable mousse should be cooked to a minimum internal temperature of 145°F.
    True   False

18. Using the traditional ratio for types of meat used in a forcemeat, the mixture should contain 2.6 pounds of venison and 1.3 pounds of pork.
    True   False

19. Smithfield and Prosciutto hams are not actually cooked before consumption.
    True   False

28F. Putting It All Together

Provide a short response for each of the questions below. These questions are designed to help you connect the bigger concepts presented in this chapter and/or text.

1. Take a look at the list of ingredients in recipe 28.3, the meat, eggs, seasonings, etc. Looking at the whole list while thinking back to the chapter on beef, what other common recipe that you know of contains similar ingredients but perhaps in difference quantities?

2. Review the paragraph in this chapter that discusses panadas. What other recipes/preparations other than forcemeats can you think of that use a panada to enhance smoothness of the final product?

29

HORS D'OEUVRE
AND CANAPÉS

➤ **TEST YOUR KNOWLEDGE**

The practice sets provided below have been designed to test your comprehension of the information found in this chapter. It is recommended that you read this chapter completely before attempting these questions.

29A. Terminology

Fill in the blank spaces with the correct definition.

1. Malassol _____

2. Canapé base _____

3. Canapé spread _____

4. Canapé garnishes _____

5. Barquette _____

6. Tartlet _____

7. Profiterole _____

8. Crudité _____

9. Sushi _____

10. Sashimi _____

11. Rumaki _____

12. Wonton skins _____

29B. Caviar

For each of the Caviars listed below, describe the following:

    a. Price

    b. Source

    c. Consistency

1. Beluga caviar _____

    a._____

    b. _____

    c._____

2. Osetra caviar _____

    a._____

    b. _____

    c._____

3. Sevruga caviar _____

    a._____

    b. _____

    c._____

4. Pressed caviar_____

    a._____

    b. _____

    c._____

5. American Sturgeon caviar _____

    a._____

    b. _____

    c._____

6. Golden Whitefish caviar _____

    a._____

    b. _____

    c._____

7. Lumpfish caviar _____

    a._____

    b. _____

    c._____

8. Salmon caviar_____

    a._____

    b. _____

    c._____

29C. Multiple Choice

For each request below choose the one correct response.

1. Caviar is best stored at:
   a. 34°F
   b. 30°F
   c. 35°F
   d. 32°F

2. Connoisseurs prefer to serve caviar in which of the following utensils?
   a. Glass
   b. China
   c. Metal
   d. Plastic

3. Which of the following fish is *not* used to make sushi?
   a. Ahi tuna
   b. Flounder
   c. Sea bass
   d. Salmon

4. Rice wine is also known as:
   a. Wasabi
   b. Mirin
   c. Shoyu
   d. Nori

5. When making brochettes as appetizers all of the following should be considered *except*:
   a. Length of the skewer should be 4 to 8 inches long
   b. Served with a dipping sauce
   c. Carefully cut main ingredients for consistent size and shape
   d. Leave enough skewer exposed so diners can pick up easily

29D. Short Answer

Provide a short response that correctly answers each of the following requests.

1. The three (3) main ingredients found in sushi are:

   a._____

   b. _____

   c._____

2. List the four (4) guidelines for preparing hors d'oeuvres.

   a._____

   b. _____

   c._____

   d. _____

3. List six (6) canapé spreads and an appropriate garnish for each.

   a._____

   b. _____

   c._____

   d. _____

   f. _____

4. Name and describe four (4) seasonings used in making sushi.

   a._____

   b. _____

   c._____

   d. _____

5. Briefly describe the six (6) guidelines for preparing appetizers.

   a._____

   b. _____

   c._____

   d. _____

   f. _____

6. List the three (3) guidelines for preparing canapé spreads.

   a._____

   b. _____

   c._____

7. Describe the three (3) factors that indicate the freshness of caviar.

   a._____

   b. _____

   c._____

29E. Fill in the Blank

Fill in the blank provided with the response that correctly completes the statement.

1. Small skewers holding a combination of meat, poultry, game, fish or vegetables are called _____ .

2. Wontons can be steamed, but are more often _____-_____or _____-_____.

3. If hors d'oeuvres are being served before a meal then _____ to _____pieces per person should be prepared for an hour cocktail reception. However, if they are being served alone with no meal following, then _____ to _____ pieces per person per hour will probably be needed.

4. The key to good sushi is the freshness of the fish, which should be no more than _____ day(s) out of the water.

29F. Matching

Match each of the ingredients in List A with the appropriate description in List B. Each item in List B can only be used once.

List A

_____ 1. Nori

_____ 2. Wasabi

_____ 3. Shoyu

List B

a. Japanese soy sauce, which is lighter and more delicate than the Chinese variety

b. Fresh ginger pickled in vinegar

c. A strong aromatic root, purchased as a green powder, sometimes called horseradish but no relation

d. A dried seaweed, used to add flavor and to contain the rolled rice and other ingredients

29G. Chapter Review

For each statement below, circle either True or False to indicate the correct answer. If an answer is false, then explain why.

1. An appetizer is usually served before lunch.
    True   False

2. Frozen caviar should only be used as a garnish.
    True   False

3. Most refrigerators are warmer than 32°F and therefore caviar should be stored on ice.

   True   False

4. Caviar should be served in a stainless steel bowl because such an implement keeps it cooler.

   True   False

5. The primary purpose of spreading the canapé base with butter is to add flavor.

   True   False

6. Canapés are best stored over night in the refrigerator before service.

   True   False

7. The best quality caviar is always the most expensive.

   True   False

8. If properly handled, fresh caviar will last up to two weeks before opening.

   True   False

9. Sushi is prepared by adding rice wine and other seasonings to long-grain rice.

   True   False

10. Fruits de Mer en Bouchée is a convenient hot hors d'oeuvre to serve because it has a long holding time.

    True   False

11. Rumaki is a hot hors d'oeuvre that uses the barding method.

    True   False

12. Regardless of how or when hors d'oeuvres are served, attractive preparation and display are vital guidelines to follow.

    True   False

13. Hot hors d'oeuvres can be displayed on trays or platters provided they are replaced frequently to ensure they stay hot.

    True   False

29H. Putting It All Together

Provide a short response for each of the questions below. These questions are designed to help you connect the bigger concepts presented in this chapter and/or text.

1. What food safety precautions does the chef need to take when preparing and serving large quantities of hot hors d'oeuvres such as stuffed, fried wontons and Rumaki for a cocktail party that will last two hours?

2. A chef is preparing a brunch buffet that will be open for a total of 4 hours on Sunday morning. She is considering including a display of sushi (previously prepared) on the buffet. What precautions should she take?

30

PRINCIPLES
OF THE BAKESHOP

➢ <u>TEST YOUR KNOWLEDGE</u>

The practice sets provided below have been designed to test your comprehension of the information found in this chapter. It is recommended that you read this chapter completely before attempting these questions.

30A. Terminology

Fill in the blank spaces with the correct definition.

1. Conching _____

2. Tapioca_____

3. All-purpose flour _____

4. Dough_____

5. Gluten_____

6. Hygroscopic _____

7. Cooked syrups _____

8. Chocolate liquor or mass _____

9. Granulated sugar _____

10. Nib _____

11. Molasses _____

12. Sugar syrup _____

13. Emulsions _____

14. Cocoa powder _____

15. Fermentation _____

16. Extracts _____

17. Turbinado sugar _____

18. Gelatinization _____

19. Sucrose _____

20. Refined/table sugar _____

21. Raw sugar _____

22. Bloom (related to gelatin) _____

23. Simple syrup _____

24. Interferents _____

25. Mixing methods _____

26. Gelatin _____

27. Starch retrogradation _____

28. Batter _____

29. Cocoa butter _____

30. Couverture _____

31. Friable _____

32. Soft flour _____

33. Strong flour _____

34. Rye flour _____

35. Cornstarch _____

36. Arrowroot _____

37. Tapioca _____

38. Almonds _____

39. Brazil nuts _____

40. Cashews _____

41. Hazelnuts _____

42. Peanuts _____

43. Walnuts _____

44. Pinenuts _____

45. Chestnuts _____

46. Coconuts _____

47. Macadamia nuts _____

48. Pecans _____

49. Pistachios _____

50. Fat bloom _____

51. Sugar bloom _____

30B. Matching

Match each of the terms in List A with the appropriate letter in List B. Each choice in List B can only be used once.

	List A		List B
_____	1. Blending	a.	Use a spoon or electric mixer with paddle attachment
_____	2. Cutting	b.	Use a whisk or electric mixer with whip attachment
_____	3. Sifting	c.	Use a rubber spatula
_____	4. Whipping	d.	Use a spoon, rubber spatula, whisk or electric mixer with paddle attachment
_____	5. Folding	e.	Use an electric mixer with paddle attachment on medium speed
_____	6. Creaming	f.	Use a rotary or drum sifter or mesh strainer
_____	7. Beating	g.	Use a whisk, spoon or rubber spatula
_____	8. Kneading	h.	Use a flat cake spatula or metal spatula
_____	9. Stirring	i.	Use pastry cutters, fingers or an electric mixer with paddle attachment
		j.	Use hands or an electric mixer with dough hook attachment

30C. Multiple Choice

For each request below, choose the one correct response.

1. All fats are considered to be shortenings in baking because they tenderize the product and:
   a. Leaven
   b. Strengthen the gluten strands
   c. Give good color
   d. Shorten the gluten strands

2. Composite flours are:
   a. Made from corn, soybeans and rice
   b. Categorized as non-wheat flours
   c. Naturally high in protein
   d. Made with the bran intact

3. Sanding sugar is primarily used for:
   a. A granulated sugar substitute
   b. Making light, tender cakes
   c. Decorating cookies and pastries
   d. Making icings and glazes for decorating

4. The most frequently used and therefore the most important ingredient in the bakeshop is:
   a. Granulated sugar
   b. Wheat flour
   c. Shortening
   d. Yeast

5. Whole wheat flour, which includes the bran and germ, is also called:
   a. Wheat germ
   b. Composite flour
   c. Whole flour
   d. Graham flour

6. Which of the following is *false* about the role of sugar and sweeteners in the bakeshop? They:
   a. Act as a crisping agent
   b. Sserve as a preservative
   c. Tenderize products
   d. Act as a creaming agent

7. A baked good's final texture is determined by the rise, which is caused by the _____, _____, and _____ in the dough or batter.
   a. Temperature, sugar, yeast,
   b. Protein, gluten, strands
   c. Glutenin, gliadin, water
   d. Carbon dioxide, air, steam

8. 160° F is the temperature at which gluten, dairy and egg proteins _____.
   a. Brown
   b. Soften
   c. Crystallize
   d. Solidify

9. Which statement is *false* about cooked sugars?
   a. As sugars caramelize their sweetening power decreases
   b. As water evaporates the temperature of the sugar rises
   c. As sugars caramelize their sweetening power increases
   d. The sugar's temperature indicates its concentration

10. A change in a baked good's texture and starch granule structure results in:
    a. Staling
    b. Browning
    c. Leavening
    d. Gluten development

11. _____ is the brown powder left after the _____ is removed.
   a. Unsweetened chocolate, sugar
   b. Cocoa powder, sugar
   c. Milk chocolate, dairy solids
   d. Cocoa powder, cocoa butter

12. Sugar is graded in the following manner:
   a. U.S. Grade No.1, No. 2, No. 3, No. 4
   b. There are no government standards regulating grade labels
   c. U.S. Grade Extra Fancy, Fancy, Good, Standard
   d. U.S. Superior Grade, Standard Grade, Good Grade

13. The purpose of a hydrometer is to:
   a. Measure specific gravity and degrees of concentration
   b. Show the temperature of sugar syrups
   c. Determine doneness
   d. Measure the amounts of sugar needed in sugar syrups

14. Which statement regarding measuring ingredients is *untrue?*
   a. Most foods do not weigh their volume
   b. Baking formulas usually use weight as the primary means of measure
   c. Spring scales are most commonly used for measuring in the bakeshop
   d. Precise, accurate measurement of ingredients is vital in the bakeshop

15. Which *two* of the following products *do not* contain chocolate products?
   a. White chocolate truffles with Grand Marnier
   b. Nestle's Toll House cookies with semisweet chocolate chips
   c. Hershey's milk chocolate bar with almonds
   d. Chocolate flavored solid chocolate Easter bunny

30D. Short Answer

Provide a short response that correctly answers each of the following requests.

1. List six (6) important steps to follow when melting chocolate.

   a. _____

   b. _____

   c. _____

   d. _____

   e. _____

   f. _____

2. List four (4) differences between unsweetened (baking) and bittersweet/semisweet chocolate.

Unsweetened	Bittersweet/Semisweet
a._____	_____
b. _____	_____
c._____	_____
d. _____	_____

3. Name three (3) characteristics achieved by tempering chocolate.

a._____

b. _____

c._____

4. List five (5) types of equipment used specifically in the bakeshop.

a._____ d. _____

b. _____ e. _____

c._____

5. What two (2) things can prevent the development of gluten in a recipe (making it more elastic).

a._____

b. _____

30E. Chapter Review

For each statement below, circle either True or False to indicate the correct answer. If an answer is false, then explain why.

1. Self-rising flour is bread flour with salt and baking powder added to it.

   True   False

2. Glutenin and gliadin contain the gluten necessary to create a quality dough or batter.

   True   False

3. Chocolate did not exist in Europe as we know it today until Columbus brought the first cacao beans back to Spain from the New World.

   True   False

4. Unsweetened chocolate is pure hardened cocoa butter.

   True   False

5. Most white chocolate products are not made from cocoa beans since they substitute vegetable oils for cocoa butter.

   True   False

6. Chocolate will melt just below body temperature.

   True   False

7. Gluten provides structure in dough by enabling the gases from fermentation to be retained.

   True   False

8. Flour derived from the portion of the endosperm closest to the germ is coarser.

   True   False

9. Whole wheat flours have a shorter shelf life due to their fat content.

   True   False

10. Unopened bags of flour can be stored anywhere as long as the location is relatively cool and free of moisture.

    True   False

11. Beets and sugar cane are the two main sources for sugar.

    True   False

12. Unsalted butter is usually preferred to salted butter in baking strictly because the salt may interfere with the product formula.

    True   False

13. The concept of carry over baking is similar to carry over cooking.

    True   False

14. A batter generally contains more fat, sugar and liquid than dough.

    True   False

15. Fats are like other bakeshop ingredients in that they will combine completely with liquids.

    True   False

16. Shortenings are versatile and therefore oils may be substituted for solid shortening, regardless of what the recipe suggests.

    True   False

17. Granulated and sheet gelatin may be used interchangeably in a formula providing the same weight of both ingredients is used.

    True   False

18. If vanilla beans develop a white coating while in storage, they have become contaminated with mold and should not be used.

    True   False

19. Milk chocolate chips may be substituted for semisweet chocolate chips when making chocolate chip cookies.

    True   False

20. All countries use similar refining processes for chocolate and therefore there is little difference in the resulting texture.

    True   False

21. The real definition of a nut is the edible, single-seed kernel of a fruit surrounded by a hard shell.

    True   False

22. There are several types of bakeshop ovens to choose from, including convection, conventional and steam injected.

True   False

30F. Putting It All Together

Provide a short response for each of the questions below. These questions are designed to help you connect the bigger concepts presented in this chapter and/or text.

1. Most everyone is familiar with the dense, chewy texture of a bagel. With your newfound knowledge regarding flour and gluten and your experiences eating bagels, without looking up a recipe try to determine what kind of flour is used in many bagel formulas.

31

QUICK BREADS

➢ TEST YOUR KNOWLEDGE

The practice sets provided below have been designed to test your comprehension of the information found in this chapter. It is recommended that you read this chapter completely before attempting these questions.

31A. Terminology

Fill in the blank spaces with the correct definition.

1. Streusel _____

2. Biscuit method _____

3. Single-acting baking powder _____

4. Creaming method _____

5. Tunneling _____

6. Muffin method _____

7. Double-acting baking powder _____

31B. Short Answer

Provide a short response that correctly answers each of the following requests.

1. Suggest a reason why muffins might have a soapy or bitter taste.

   _____

2. Why might a recipe call for both baking soda and baking powder?

   _____

   _____

3. What situation might call for the use of double acting baking powder?

_____

4. What does the higher fat content in the creaming method do to the gluten in the mixture and therefore for the final product?

_____

_____

5. Explain why the fat is softened in recipes using the creaming method?

_____

_____

6. Suggest a reason for elongated holes running through the center of baked muffins.

_____

7. What is the basic difference between a scone and a biscuit?

_____

_____

8. List three (3) products that result from using the biscuit method of mixing.

a._____

b. _____

c._____

31C. Chapter Review

For each statement below, circle either True or False to indicate the correct answer. If an answer is false, then explain why.

1. Bread flour is used to make biscuits.

   True   False

2. The creaming method is comparable to the mixing method.

   True   False

3. Honey, molasses, fresh fruit and buttermilk are all examples of acids that may be used with baking soda.

   True   False

4. Baking powder requires an acid ingredient in the formula in order to create the chemical reaction.

   True   False

5. Some quick breads use yeast as the leavening agent.

   True   False

6. Too much kneading toughens biscuits.

   True   False

7. Fats used in the muffin method should be in a solid form.

    True   False

8. The reason for a flat top on a loaf of banana bread is probably that the leavening agent was not sufficiently strong.

    True   False

9. When carbon dioxide is trapped within a batter or dough it expands when heated, causing the product to rise.

    True   False

10. Batters and doughs made with single-acting baking powder do not need to be baked immediately, as long as the product is refrigerated immediately.

    True   False

11. In order for baking soda to leaven, a batter or dough must be baked.

    True   False

12. Shortcakes are made using the muffin method.

    True   False

13. The biscuit method of mixing results in flaky, tender products particularly due to the way the chilled solid fat is cut into the other ingredients.

    True   False

14. A muffin batter has a different composition from a quick bread batter and therefore must be baked in a different way.

    True   False

15. Liquid fat and the general mixing procedure used in the muffin mixing method generally results in a cake-like baked product.

    True   False

31D. Putting It All Together

Provide a short response for each of the questions below. These questions are designed to help you connect the bigger concepts presented in this chapter and/or text.

1. Turn to recipe 31.2 and study it. At the end of the recipe the author makes some recommendations for substituting ingredients that flavor the muffin. Why do you think the pecan spice muffin calls for only 4 ounces of pecans while the blueberry muffin recipe calls for 5 ounces?

2. Think back to the last time you had a dessert with streusel topping and how crumbly it was. Now look at recipe 31.4, study the ingredients and based on your observations take an educated guess of why it has that texture either by the ingredients listed or the ratio in which they exist in the recipe.

32

YEAST BREADS

➤ TEST YOUR KNOWLEDGE

The practice sets provided below have been designed to test your comprehension of the information found in this chapter. It is recommended that you read this chapter completely before attempting these questions.

32A. Terminology

Fill in the blank spaces with the correct definition.

1. Fermentation _____

2. Punching down _____

3. Straight dough method _____

4. Oven spring _____

5. Slashing or docking _____

6. Fresh yeast _____

7. Wash _____

8. Sponge method _____

9. Proof box _____

10. Rolled-in doughs _____

11. Rounding _____

12. Proofing _____

13. Quick-rise dry yeast _____

14. Absorb (in relation to scaling) _____

15. Bulk fermentation _____

16. Couche _____

17. Bannetons _____

18. Brotform _____

32B. Multiple Choice

For each request below, choose the one correct response.

1. Which is an example of a rich dough?
   a. Biscuits
   b. Italian bread
   c. Challah bread
   d. Muffins

2. Quick-rise dry yeast uses _____ water in order to activate the fermentation process.
   a. 138°F
   b. 95° F
   c. 100 to 110°F
   d. 125 to 130°F

3. When yeast is combined with carbohydrates, the result is alcohol and:
   a. Oxygen
   b. Gas
   c. Carbon dioxide
   d. Water

4. The disadvantage of using butter in rolled-in doughs is that it:
   a. Has a high moisture content
   b. Cracks and breaks
   c. Adds too much salt to the dough
   d. Needs to be clarified before using

5. Yeast products should be cooled to approximately what temperature?
   a. 32 to 34°F
   b. 60 to 70°F
   c. 45 to 50°F
   d. 80 to 90° F

6. Which of the following is *not* important when considering the amount of flour used in a yeast bread?
   a. Percentage of salt in the formula
   b. Flour storage conditions
   c. Humidity level
   d. Measuring accuracy of other ingredients

7. Commercial baking yeast was not made available in stores until:
   a. 1654
   b. 1857
   c. 1910
   d. 1868

8. There are primarily two market forms of bakers' yeast. They are:
    a. Compressed and active dry
    b. Brewers and compressed
    c. Quick-rise dry and instant
    d. Fresh and compressed

9. The primary chemical function of rounding is to:
    a. Smooth the dough into round balls
    b. Stretch the gluten into a smooth coating
    c. Help retain the gases from fermentation
    d. Proof the dough

10. When is "punching" performed?
    a. After initial fermentation
    b. After proofing
    c. During proofing
    d. After initial mixing of dough

32C. Short Answer

Provide a short response that correctly answers the following requests.

1. Explain the two (2) steps involved in the sponge method.

    a. _____

    b. _____

2. Why is the organism in active dry yeast considered dormant?

    _____

3. List four (4) factors for determining doneness of a baked yeast-leavened product.

    a. _____

    b. _____

    c. _____

    d. _____

4. List three (3) examples of a rolled-in dough product.

    a. _____

    b. _____

    c. _____

5. How is the quantity of dry yeast determined when it is being substituted for compressed yeast?

    _____

6. Describe the method for producing a straight method dough.

    _____

7. Briefly list the ten (10) sequential stages of yeast bread production.

a. _____   f. _____

b. _____   g. _____

c. _____   h. _____

d. _____   i. _____

e. _____   j. _____

8. Visually examining bread can be used as one method for determining doneness, but for the less experienced, taking an internal temperature is more accurate. Using an instant read thermometer, provide the internal temperatures bread should reach before it is cooked throughout:

Lean bread dough: _____

Rich bread dough: _____

32D. Chapter Review

For each statement below, circle either True or False to indicate the correct answer. If an answer is false, then explain why.

1. Punching down occurs before the proofing process.
   True   False

2. Salt's primary role in bread making is seasoning.
   True   False

3. Italian bread is an example of a product made using the straight dough method.
   True   False

4. Washes can be applied before or after proofing occurs.
   True   False

5. Underproofing may result in a sour taste, poor volume and a paler color after baking.
   True   False

6. Rich doughs are baked without steam.
   True   False

7. When properly stored, compressed yeast has a shelf life of two to three weeks.
   True   False

8. Active dry yeast contains approximately 10% moisture content.
   True   False

9. Instant yeast can be substituted measure for measure for regular dry yeast.
   True   False

10. Starters are used primarily for flavor in bread making.
    True   False

11. There is very little difference between the flavors of dry and compressed yeasts.
    True   False

12. Overkneading is a common problem in bread making.

    True   False

13. Yeast is very sensitive to temperature and moisture and prefers temperatures between 60°F and 90°F.

    True   False

14. To prepare compressed yeast for using in a formula, soften it in twice its weight in warm water at 100°F before adding it to bread dough.

    True   False

15. Yeast is a living organism and instant dry yeast will be destroyed at temperatures above 138°F.

    True   False

16. Other yeasts used in baking, aside from instant dry yeast, are destroyed at temperatures above 138°F.

    True   False

17. The fermentation process happens in only a few hours. If a bakery wants to slow down that process and extend it over 8-10 hours (or more) so the flavors of the bread become more complex, such as in artisan-style breads, they can put the dough into a piece of equipment called a retarder.

    True   False

32E. Putting It All Together

Provide a short response for each of the questions below. These questions are designed to help you connect the bigger concepts presented in this chapter and/or text.

1. Can someone allergic to wheat still eat bread? Answer the question based on your knowledge of various types of flours.

2. In layman's terms explain how a bread dough rises. Use your knowledge of yeast, flour, sugar and a liquid to explain how the process works.

33

PIES, PASTRIES
AND COOKIES

➤ **TEST YOUR KNOWLEDGE**

The practice sets below have been designed to test your comprehension of the information found in this chapter. It is recommended that you read this chapter completely before attempting these questions.

33A. Terminology

Fill in the blank spaces with the correct definition.

1. Flaky dough _____

2. Pâte Sucrée _____

3. Meringue _____

4. Pâte à choux_____

5. Soft meringue _____

6. Paris-Brest_____

7. Wafer cookies _____

8. Cream filling _____

9. Mealy dough _____

10. Éclair paste _____

11. Italian meringue_____

12. Cream puffs _____

13. Bouchées _____

14. Sweet dough _____

15. Pie _____

16. Churros _____

17. Bar cookies _____

18. Hard meringue _____

19. Custard filling _____

20. Chiffon _____

21. Tart _____

22. Rolled/cut out cookies _____

23. Vol-au-vents _____

24. Crumbs _____

25. Common meringue _____

26. Puff pastry _____

27. Icebox cookies _____

28. Fruit fillings: cooked, cooked juice, baked _____

29. Detrempe _____

30. Swiss meringue _____

31. Baked blind _____

32. Pâte feuilletée _____

33. Crullers _____

34. Drop cookies _____

35. Beignets _____

36. Pressed cookies _____

37. Éclairs _____

38. Feuilletées _____

33B. Short Answer

Provide a short response that correctly answers the following requests.

1. List three (3) types of fillings that are used to fill prebaked pie crusts.

    a. _____   c. _____

    b. _____

2. List four (4) types of fillings that are appropriate for filling a crumb crust.

    a. _____   c. _____

    b. _____   d. _____

3. What two (2) types of fillings are cooked by baking them *in* a crust?

   a. _____   b._____

4. List three (3) reasons for using a flaky dough to prepare pies.

   a. _____   c. _____

   b. _____

5. Why is a sweet dough, or pâte sucrée, better for making tarts?

   _____

6. When is it appropriate to use a mealy crust?

   _____

7. Why is hand mixing best when making small to moderate quantities of flaky dough?

   _____

   _____

8. What makes pâte à choux unique among doughs?

   _____

   _____

9. What determines whether a meringue is hard or soft?

   _____

10. List four (4) uses for puff pastry.

    a._____

    c._____

    b. _____

    d. _____

11. List three (3) factors that have an effect on a cookie's texture.

    a._____

    c._____

    b. _____

33C. Multiple Choice

For each request below, choose the one correct response.

1. What is the most common method for preparing cookie doughs?
   a. Beating
   b. Whipping
   c. Blending
   d. Creaming

2. Which is *not* a use for pâte à choux?

   a. Profiteroles

   b. Palmiers

   c. Éclairs

   d. Paris-Brest

3. What do all meringues have in common?

   a. The ratio of egg whites to sugar

   b. Whipped egg whites and sugar

   c. The flavoring ingredient used

   d. The method of preparation

4. Lacy pecan cookies are a cookie variety classified as a(n):

   a. Pressed cookie

   b. Icebox cookie

   c. Wafer cookie

   d. Drop cookie

5. Egg whites will whip better if _____ before whipping:

   a. A small amount of salt is added

   b. They are well chilled

   c. A portion of the sugar is added

   d. They are brought to room temperature

33D. Chapter Review

For each statement below, circle either True or False to indicate the correct answer. If an answer is false, then explain why.

1. A baked meringue containing ground nuts is a dacquoise.

   True   False

2. Cherries and apples are appropriate fruits to use for a cooked juice filling.

   True   False

3. A cream filling is basically a flavored pastry cream.

   True   False

4. Pumpkin pie is a good example of a custard filling.

   True   False

5. The ratio for making a crumb crust is one part sugar, to four parts crumbs, to two parts melted butter.

   True   False

6. Rice or beans can be used for blind baking.

   True   False

7. Any dough can be used to make a tart shell as long as it tastes good and has a good appearance.

   True   False

8. Italian and Swiss meringues work equally well in buttercreams.

   True   False

9. A cooked juice filling should be combined with a raw crust and then baked.

   True   False

10. Baked fruit pies must be refrigerated to retard bacterial growth.

    True   False

11. The type of fat used in a flaky or mealy dough affects both the flavor and flakiness.

    True   False

12. An American gâteau is a pastry item made with puff pastry, éclair paste, short dough or sweet dough.

    True   False

33E. Putting It All Together

Provide a short response for each of the questions below. These questions are designed to help you connect the bigger concepts presented in this chapter and/or text.

1. Based on your knowledge of fats and flour, what might happen to a blind baked pie shell if stored at room temperature (wrapped) for more than 2-3 days?

2. What food safety issue may exist for cream filled and custard pies and what does that issue indicate about shelf life?

34

CAKES AND FROSTINGS

> ## TEST YOUR KNOWLEDGE

The practice sets provided below have been designed to test your comprehension of the information found in this chapter. It is recommended that you read this chapter completely before attempting these questions.

34A. Terminology

Fill in the blank spaces with the correct definition.

1. Creamed fat _____

2. Whipping eggs _____

3. Butter cakes _____

4. Creaming method cakes _____

5. High-ratio cakes _____

6. Genoise _____

7. Spongecakes _____

8. Angel food cakes _____

9. Chiffon cakes _____

10. Icing _____

11. Buttercream _____

12. Foam _____

13. Fondant _____

14. Glaze _____

15. Royal icing _____

16. Ganache _____

17. Simple buttercream _____

18. Italian buttercream _____

19. French buttercream _____

20. Meringue buttercream _____

21. Mousseline buttercream _____

22. Boiled icing _____

23. Glucose _____

24. Flat frosting _____

25. Decorator's icing _____

26. Side masking _____

27. Stencils_____

28. Baker's comb _____

29. Two-stage mixing method _____

34B. Basic Cake Mixes

Describe the *basic* steps for the preparation of the following cake mixes and give a menu example of each type of cake. Exact quantities of ingredients are not necessary for this exercise.

Example: Chiffon Cake

    a.   Whip egg whites with a little sugar until stiff.

    b.   Add liquid ingredients, including oil, to sifted dry ingredients.

    c.   Fold in egg whites.

    d.   Bake in ungreased pan.

Menu example: Lemon Chiffon Cake

1.   Butter Cake:

    _____

    _____

    _____

    _____

    _____

    Menu example:_____

2. Genoise Cake:

_____

_____

_____

_____

_____

Menu example:_____

3. Sponge Cake:

_____

_____

_____

_____

_____

Menu example:_____

4. Angel Food Cake:

_____

_____

_____

_____

_____

Menu example:_____

34C. Matching I Ingredient Functions

Match each of the classification headings in List A with the appropriate ingredients in List B.
Each choice in List B can only be used once.

List A

_____ 1. Flavoring
_____ 2. Toughener
_____ 3. Leavener
_____ 4. Tenderizer
_____ 5. Drier
_____ 6. Moistener

List B

a. Flour, milk, eggs
b. Flour, butter, water
c. Sugar, fats, yolks
d. Flour, starches, milk solids
e. Baking powder, baking soda
f. Cocoa, chocolate, spices, sour cream
g. Water, milk, juice, eggs

34D. Matching II Frostings

Match each of the frostings in List A with the appropriate ingredients in List B. Each choice in List B can only be used once.

List A

_____ 1. Ganache

_____ 2. Fudge

_____ 3. Royal icing

_____ 4. Glaze

_____ 5. Buttercream

_____ 6. Fondant

List B

a. Sugar, fat, egg yolks or whites

b. Uncooked confectioner's sugar and egg whites

c. Meringue made with hot sugar syrup

d. Blend of melted chocolate and cream

e. Confectioner's sugar with liquid

f. Cooked mixture of sugar, butter and water/milk; applied warm

g. Cooked mixture of sugar, and water, applied warm

34E. Cake Mixing Categories

Cake mixes fall into one of two categories: creamed fat or whipped egg. From the list below identify the creamed fat mixes with the letter A and the whipped egg mixes with the letter B.

A = Creamed Fat

B = Whipped Egg

_____ 1. Chiffon cake

_____ 2. Continental brownies

_____ 3. Devil's food cake

_____ 4. Chocolate sponge cake

_____ 5. Sacher torte

_____ 6. Yellow cake

_____ 7. Carrot cake

_____ 8. Gateau Benoit

_____ 9. Vanilla raspberry layer cake

_____ 10. Sour cream coffee cake

34F. Short Answer—Frostings

Describe the *basic* steps for the preparation of the following frostings. Exact quantities are not important for this exercise.

1. Simple Buttercream

_____

_____

_____

_____

_____

_____

2. Italian Buttercream

_____

_____

_____ –

_____

_____

_____

3. French Buttercream

_____

_____

_____ –

_____

_____

_____

34G. Fill in the Blank

Fill in the blank with the response that correctly completes the statement.

1. The amount of leavening should be _____ at higher altitudes and the eggs in the mixture should be _____. Temperatures should also be _____ by _____ degrees F at altitudes over 3,500 feet.

2. The three tests for determining a cake's doneness are _____, _____ and _____.

3. Most cakes are baked at temperatures between _____ degrees F and _____ degrees F.

4. Royal icing is also known as _____ _____.

5. Pan coating consists of equal parts _____, _____ and _____.

6. When mixing any cake batter, the goals are _____, _____ and _____.

34H. Chapter Review

For each statement below, circle True or False to indicate the correct answer. If an answer is false, then explain why.

1. The best way to cool a cake is to leave it in an area where there is a cool breeze.

   True   False

2. Frostings are not usually frozen.

   True   False

3. As a general guide for setting oven temperatures for cakes, the greater the surface area, the higher the temperature.

   True   False

4. When baking cakes, pans should be filled 1/2 to 3/4 with cake mix for the best results.

   True   False

5. Angel food cake is ideal for frosting.

   True   False

6. Solid shortening is better than butter for coating pans, since it does not contain any water.

   True   False

7. Package mixes are inferior in quality to cakes that are made from scratch.

   True   False

8. The fat used in high-ratio cake mixes can be either butter or shortening.

   True   False

9. When testing the doneness of a cake by touch, it should spring back quickly without feeling soggy or leaving an indentation.

   True   False

10. Compotes, buttercream, foam, fudge and ganache are all examples of icings.

    True   False

34I.   Putting It All Together

Provide a short response for each of the questions below. These questions are designed to help you connect the bigger concepts presented in this chapter and/or text.

1. When it comes down to making cakes, what should chefs watch out for with food safety?

2. How should a food service establishment determine whether to bake and decorate its own cakes/desserts versus outsourcing?

35

CUSTARDS, CREAMS, FROZEN DESSERTS AND DESSERT SAUCES

> ➤ **TEST YOUR KNOWLEDGE**

The practice sets provided below have been designed to test your comprehension of the information found in this chapter. It is recommended that you read this chapter completely before attempting these questions.

35A. Terminology

Fill in the blank spaces with the correct definition.

1. Custard _____

2. Stirred custard _____

3. Baked custard _____

4. Vanilla custard sauce _____

5. Pastry cream _____

6. Sabayon _____

7. Crème brûlée _____

8. Mousseline _____

9. Crème chibouts _____

10. Temper _____

11. Crème caramel _____

12. Cheesecake _____

13. Bread pudding _____

14. Steep _____

15. Bavarian creams _____

16. Chiffons _____

17. Mousses _____

18. Crèmes _____

19. Chantilly _____

20. Charlotte _____

21. Ice cream _____

22. Gelato _____

23. Sorbets _____

24. Sherbets _____

25. Semifreddi _____

26. Overrun _____

27. Sundae _____

28. Baked Alaska _____

29. Bombes _____

30. Coupes _____

31. Parfaits _____

32. Marquis _____

33. Neapolitans _____

34. Coulis _____

35. Base _____

36. Filling _____

37. Garnish _____

35B. Short Answer

Provide a short response that correctly answers the following requests.

1. Eggs are a high protein food and can easily become contaminated. List and describe six (6) sanitary guidelines for handling eggs.

   a. _____

   b. _____

   c. _____

   d. _____

   e. _____

   f. _____

2. Describe the basic steps and essential ingredients for the preparation of vanilla custard sauce.

_____

_____

_____

_____

_____

_____

3. Briefly describe the eight (8) sequential steps in the procedure for making ice cream.

a._____

b. _____

c._____

d. _____

e._____

f._____

g. _____

h. _____

4. Describe the four (4) sequential steps in the procedure for making baked soufflés.

a._____

b. _____

c._____

d. _____

5. Describe the four (4) sequential steps in the procedure for making a sabayon.

a._____

b. _____

c._____

d. _____

6. Describe the three (3) sequential steps in the procedure for making a mousse.

a._____

b. _____

c._____

7. Briefly describe the nine (9) guidelines for assembling desserts.

a._____

b. _____

c._____

d. _____

e. _____

f. _____

g. _____

h. _____

i. _____

8. Because ice cream is a potentially hazardous food product as well as being ready-to-eat, list three (3) precautions one can take to maintain a safe product for consumption.

a. _____

b. _____

c. _____

35C. Fill in the Blank

Fill in the blank with the response that correctly completes the statement.

1. _____ is the Italian name for sabayon.

2. Pastry cream can be lightened by folding in whipped cream to produce a _____ or by adding _____ _____ to produce a crème Chiboust.

3. Some creams such as _____ and _____ are thickened with gelatin, but others such as _____ and _____ _____ are not, and therefore are softer and lighter.

4. When preparing a soufflé the custard base and egg whites should be at room temperature because _____ and _____.

35D. Chapter Review

For each statement below, circle True or False to indicate the correct answer. If an answer is false, then explain why.

1. Once a coulis has been cooked it becomes a compote.
   True   False

2. A frozen soufflé is not really a soufflé in the true sense.
   True   False

3. Crème Anglaise is typically vanilla but may be flavored in a variety of ways.
   True   False

4. The purpose of serving a dessert with a sauce is to add moisture, texture, flavor and to enhance the presentation.
   True   False

5. A coulis is a fruit purée made from either fresh or individually quick frozen (IQF) fruits.
   True   False

6. Quiche is an example of a baked custard.
   True   False

7. The chef has made a fresh sorbet and has allowed ample time for it to freeze thoroughly, yet it is still soft and syrupy. The problem is likely that he used too many egg whites in the preparation.

True   False

8. Once a vanilla custard sauce is curdled it should be discarded.

True   False

9. A sherbet differs from a sorbet in that a sorbet contains milk or egg yolk for added creaminess.

True   False

10. A fudge sauce is a variation of a ganache.

True   False

35E. Putting It All Together

Provide a short response for each of the questions below. These questions are designed to help you connect the bigger concepts presented in this chapter and/or text.

1. In the section on souffles the formula provided (34.6) asks the chef to "fold the whipped egg whites into the other ingredients." Create a description that compares the differences in the processes of mixing versus folding.

2. In table 35.1 the reader is told that gelatin is an optional ingredient when making a mousse. Since it is left up to the chef to decide, what characteristic does gelatin add to the formula that would help him/her to decide whether to use it or not?

3. Quality ice creams, sherbets and sorbets are all available from retailers. What factors might one consider when deciding to purchase these as convenience products versus making them in-house?

36

PLATE PRESENTATION

➢ **TEST YOUR KNOWLEDGE**

The practice sets provided below have been designed to test your comprehension of the information found in this chapter. It is recommended that you read this chapter completely before attempting these questions.

36A. Terminology

Fill in the blank spaces with the correct definition.

1. Service _____

2. Composition_____

3. Presentation _____

36B. Fill in the Blank

Fill in the blanks provided with the response that correctly answers each statement.

1. _____ is a cookielike dough piped very thin and baked for use in making decorations and garnishes.

2. Proper cooking procedures can enhance the _____, _____ and _____ of many cooked foods.

3. Once the color or pattern is chosen for a plate, the next important element to consider is _____, keeping in mind the amount of food being presented.

4. The _____ is often the highest point on the plate.

5. Plate drawing is most typically done with _____ sauces.

36C. Short Answer

Provide a short response that correctly answers the following requests.

1. List two (2) ways of presenting polenta.

    a. _____

    b. _____

2. List five (5) things that a sauce should add to a plate presentation.

    a._____   d. _____

    b. _____   e. _____

    c._____

3. List two (2) things that a hippen masse garnish might do for a presentation.

    a. _____

    b. _____

4. List three (3) reasons for carefully cutting foods.

    a. _____

    b. _____

    c. _____

5. List three (3) guidelines for arranging foods on a plate.

    a. _____

    b. _____

    c. _____

6. List three (3) reasons for molding foods.

    a. _____

    b. _____

    c. _____

36D. Chapter Review

For each statement below, circle either True or False to indicate the correct answer. If an answer is false, then explain why.

1. A rice pilaf is a good example of a dish that would mold well for an attractive presentation.
    True   False

2. The garnish should always be the focal point of the plate.
    True   False

3. Generally speaking foods with similar textures look boring together.
    True   False

4. Dusting a plate can be done after the plating of the food is complete.

   True   False

5. A piping bag would be a good choice of equipment for performing sauce drawings.

   True   False

6. The primary consideration with sauce drawing is that the colors of the sauces used contrast each other.

   True   False

7. Properly preparing the main food product on a plate is the most important way to make the presentation look attractive.

   True   False

8. Plate dusting is a garnishing technique most associated with culinary preparations.

   True   False

9. Presenting food on a plate that is proportionately too large may make the food portion look too sparse thereby creating poor value perception.

   True   False

10. The most important factor to consider when decorating a plate presentation with simple or elaborate details is cost.

    True   False

36E. Putting It All Together

Provide a short response for each of the questions below. These questions are designed to help you connect the bigger concepts presented in this chapter and/or text.

1. This chapter discusses using two large spoons to form purees or mousses into a quenelle shape. What things does a chef have to ensure so that this plan is carried out effectively when it comes to the time to actually plate the food?

2. Considering the time/temperature principle learned in Chapter Two and the need to serve foods at the appropriate temperature, what rules does a chef need to adopt as a focus when planning food presentations?

37

BUFFET PRESENTATION

> ## TEST YOUR KNOWLEDGE

The practice sets provided below have been designed to test your comprehension of the information found in this chapter. It is recommended that you read this chapter completely before attempting these questions.

37A. Terminology

Fill in the blank spaces with the correct definition.

1. Theme _____

2. Menu _____

3. Chafing dish _____

4. Grosse piece _____

5. Risers _____

6. Butler service _____

37B. Short Answer

Provide a short response that correctly answers each of the following requests.

1. Describe the four (4) guidelines for avoiding repetition on a buffet.

    a._____

    b. _____

    c._____

    d. _____

2. Explain why each of the following aspects of presentation are important when preparing a buffet.

   a. Height: _____

   _____

   b. Pattern: _____

   _____

   c. Color: _____

   _____

   d. Texture: _____

   _____

   e. Negative space: _____

   _____

3. Describe three (3) styles of buffet set-up designed to promote efficient service of large groups.

   a. _____

   b. _____

   c. _____

4. Describe four (4) guidelines for presenting *hot* foods on a buffet.

   a. _____

   b. _____

   c. _____

   d. _____

37C. Multiple Choice

For each request below, choose the one correct response.

1. On a buffet, *hot* foods are served in/on:
   a. Platters
   b. Bowls
   c. Chafing dishes
   d. Mirrors

2. Which of the following best describes the logical flow for a buffet table?
   a. Appetizers, entrees, plates, vegetable
   b. Plates, soups, entrees, desserts
   c. Plates, entrees, vegetable, appetizers
   d. Desserts, vegetable, entrees, appetizers

3. Which of the following items are appropriate to use as a centerpiece on a buffet table?

    a. Flowers

    b. Ice carvings

    c. Whole turkey

    d. All of the above

4. Having waiters stationed behind the buffet table is called:

    a. Waiter service

    b. Buffet service

    c. Butler service

    d. Restaurant service

5. Below are several measures that can be taken to prevent food safety issues on a buffet *except*:

    a. provide clean serving utensils regularly

    b. don't add new food to old food in a serving dish

    c. reheat food in the chafing dish

    d. provide an ample supply of clean plates

37D. Chapter Review

For each statement below circle either True or False to indicate the correct answer. If an answer is false, then explain why.

1. More guests help themselves to little portions of every item on a buffet rather than gorging themselves on one or two items.

    True   False

2. Cost is *not* a factor when preparing buffet menus.

    True   False

3. When designing a buffet, the food tables should be located as far away from the kitchen as possible for ease of service.

    True   False

4. The start of the buffet should be located near the entrance to the room.

    True   False

5. A buffet table with 10 items should measure approximately 10 feet long.

    True   False

6. Dishes containing sauces should be positioned at the back of the table.

    True   False

7. Dead space on the buffet should never be filled in with decorations or props.

    True   False

8. Guests will generally take larger portions of foods at the beginning of the buffet than at the end.

    True   False

9. Spaghetti is an appropriate food to serve on a hot buffet.

    True   False

10. Dishes of food on the buffet should be replenished when they are two-thirds empty.

    True   False

11. The word buffet is used to describe the event as well as the table on which the food is served.

    True   False

12. A chef should not plan one portion of each food item on the buffet for each guest.

    True   False

37E. Putting It All Together

Provide a short response for each of the questions below. These questions are designed to help you connect the bigger concepts presented in this chapter and/or text.

1. Referring to the sidebar in this chapter, entitled "The Buffet: An Interesting History," hypothesize (researching if necessary) how the food safety practices have likely evolved over time with the development of the buffet. Compare and contrast how the Italian banchetto may have differed during the Renaissance period compared to an American buffet in a fine hotel today.

2. Once the buffet is set up, provide a minimum of three controls the food service establishment management can do to control cross contamination of the food during its service.

ANSWER KEY

> ## CHAPTER 1 ANSWER KEY

1A. Terminology

Answers will not be provided in answer key. All answers can be found in text.

1B. Fill in the Blank

1. Back of the house
2. French
3. Front of the house
4. Russian
5. Front waiter
6. American
7. Backwaiter or busperson
8. Chef de vin
9. Internet

1C. Short Answer

1. a. Simultaneous cooking of many items, especially those needing constant and delicate attention
    b. Cooks could more comfortably and safely approach the heat source and control its temperatures
    c. Cooks could efficiently prepare and hold a multitude of smaller amounts of items requiring different cooking methods or ingredients for later use or service
2. a. Canned foods
    b. Refrigerators
    c. Freezers

    d. Freeze-drying

    e. Vacuum-packing

    f. Irradiation

3. a. Absorbs facial perspiration

    b. Disguise stains

    c. Double-breasted design can be used to hide dirt and protects from scalds and burns

    d. Protects uniform and insulates body

4. a. Personal performance and behavior

    b. Good grooming practices

    c. Clean, pressed uniform

1D. Defining Professionalism

1. Judgment

2. Dedication

3. Taste

4. Pride

5. Skill

6. Knowledge

1E. Noteworthy Chefs

1. g		4. d	
2. f		5. a	
3. c		6. e	

1F. Matching

1. a		7. b	
2. I		8. f	
3. e		9. h	
4. c		10. k	
5. g		11. d	
6. l		12. m	

1G. Chapter Review

1. True

2. False   (p. 11)   Today most food service operations utilize a simplified version of Escoffier's kitchen brigade.

3. False   (p. 10)   Most new concerns that affect the food service industry are brought on by the demands of the customer. Such concerns may eventually encourage government interaction to ensure public well-being.

4. False	(p. 9)	Advances in the transportation industry began to positively influence the food service industry during the early 1800's.
5. True		
6. False	(p. 12)	Restaurants offering buffet service generally charge by the meal; if they charge by the dish they are known as cafeterias.
7. True		
8. True		
9. True		
10. False	(p.8)	Many of the preserving techniques used prior to the 19th century destroyed or distorted the appearance and flavor of the foods. Therefore, when new preserving techniques were developed in the early 19th century, many of them were adopted due to their minimal effect on appearance and flavor.
11. True		
12. False	(p.14)	Dining in a linear fashion does not allow for a full, simultaneous satisfaction of the groups of taste which, from a Western perspective, generally do not include spicy.
13. False	(p. 7)	The book is an astounding collection of more than 5000 classic cuisine recipes and garnishes and he emphasizes the mastery of techniques, the thorough understanding of cooking, principles and the appreciation of ingredients; attributes he considered to be the building blocks professional chefs should use to create great dishes.

1H. Putting It All Together

1. Answers may vary, but here are some major ideas to consider: Understanding the history of how the culinary profession evolved enables us to appreciate the significance of our industry today. Examples may include but are not limited to: What challenges the early chefs faced, creating a profession from nothing (and realizing that such a practice can take hundreds of years), changing the way the public thinks about food for more than sustenance but rather enjoyment, preparing foods in a way people had literally never experienced before, developing early sanitation and safety practices, inventing the recipes, designing the professional persona of the chef through uniform design and modeling professional and appropriate behaviors, determining what positions are needed in a classical kitchen, developing a training system (albeit not necessarily uniform) for various positions in a professional kitchen, creating an organizational system for a classical kitchen that clearly delineates the responsibilities of all employees, defining and perfecting various cooking techniques and methods that are still used today, creating various cooking styles based on cultural influences and regional variations, and constantly redefining cuisine—taking it beyond cultural boundaries and evolving as our economy has become globalized.

2. LeNôtre, whose impact on the chef profession is more recent than others mentioned in this chapter, made tremendous advances in baking and pastry arts. The development of baking and pastry arts as we know them today began much later than culinary art and so he built on techniques developed earlier by such greats as Antonin Carême, but in the nouvelle cuisine style. In addition, unlike Carême, who was chef to a French Diplomat, and Escoffier, who developed his culinary skills in the kitchens of the finest hotels in Europe, LeNôtre was more like Fernand Point in that he owned his own establishment. Like most great chefs, he built on

the techniques and recipes developed before him, but in LeNôtre's case, he took it several steps further. He not only opened one establishment, but numerous bakeries, and not all in the same town, let alone country. In addition, he realized with the growth in his business the need to be able to train many employees in the consistent methods of food preparation so that he could ensure that the quality of products was the same, whichever one of his establishment his customers frequented. This private school for his own chefs was eventually opened up (for a fee) to competitors' chefs who also needed training. Like his predecessors he, too, helped the evolution of baking and pastry arts relative to the period of time in which he was living, developing a new mixing technique and perfecting Bavarians, Charlottes, fruit mousses, cakes and tortes. He also applied modern technologies of his day, mastering the technique of freezing as a preserving method. Some gastronomes say he single-handedly saved the classical pastry profession in the twentieth century from being taken over by mass production bakeries. The era in which he lived likely had a significant impact on his success and the impact he had on the food service profession based on the items summarized here; he was innovative and professional, benefiting from the information created in our industry by chefs who came before so that his impact became much more monumental.

3. The developments in the production of new food ingredients for use in the modern food industry have all occurred due to the demands of the day. Some have been more controversial than others. While many chefs disdain commercial produce that has been raised with the use of chemical fertilizers and pesticides, such practices have enabled farmers to better meet the food supply for a growing world population. Likewise, animal husbandry has helped produce animals that yield a better quality and proportion of meat, increasing the yield for the farmer and therefore, the chef. Aquaculture developments have made certain fish species more available in a time of over fishing and increased pollution of our rivers, lakes, streams and oceans. While some chefs challenge the resulting flavor and quality of farmed fish, the fact is that the development of the process has resulted in a more readily available food supply, often with equal or greater raw product quality, and many times fresher and safer to consume from a food safety standpoint. While these are just a few examples of how one can justify such developments, it brings up an important challenge to the modern day chef that is in line with the authors' emphasis for chefs to continue to expand their knowledge far past their culinary school education. Chefs must continually research current trends and availability of food ingredients and based on their own values and the demands of their customers, choose those that are in alignment with both philosophies.

➢ CHAPTER 2 ANSWER KEY

2A. Terminology

Answers will not be provided in answer key. All answers can be found in text.

2B. Multiple Choice

1. d   4. d
2. d   5. c
3. b   6. d

2C. Chapter Review

1. True

2. False   (p. 32)   A licensed pest control operator should be contacted immediately. Such professionals will go beyond simply locating the source of infestation, but will also prescribe a plan of action to prevent ongoing occurrences in the future.

3. False   (p. 21)   Toxins cannot be smelled, seen or tasted.

4. False   (p. 25)   Semisolid foods should be placed in containers that are less than two inches deep since increasing the surface area decreases the cooling time.

5. True

6. True

7. True

8. False   (p. 29)   Just as hands should be washed regularly during food production to prevent cross contamination, so too, should gloves be changed regularly. Wearing gloves does not eliminate the need to wash hands regularly.

9. False   (p. 26)   Hepatitis A is a virus.

10. True

11. False   (p. 25)   A low water activity level only halts bacterial growth, it does not kill the microorganisms.

12. True

13. True

14. False   (p. 29)   The two tasting spoon method involves using the first spoon to remove a sampling of the food from the pan in which it was made or stored. Pour that food into the second spoon before tasting it. The cook can repeat the process if necessary, depending on how many tastings are needed to make final adjustments to the food product, preventing the soiled spoon from going back into the food being prepared.

2D. HACCP Overview

1.   1a   4e
      6b   7f
      3c   5g
      2d   8h

2. b

3. b

4. b, c

2E. Food-Borne Diseases Review

1. Botulism

   O: Clostridium botulinum

   F: Toxins, cells, spores

   S: Cooked foods held for an extended time at warm temperature with limited oxygen, rice, potatoes, smoked fish, canned vegetables

   P: Keep internal temperature of cooked foods above 140 degrees F or below 40 degrees F; reheat leftovers thoroughly; discard swollen cans

2. Hepatitis A

   O: Virus

   S: Enters food supply through shellfish harvested from polluted waters. It is also carried by humans, often without knowledge of infection.

   P: Confirm source of shellfish, good personal hygiene, avoid cross-contamination.

3. Strep

   O: Streptococcus

   F: Cells

   S: Infected food workers

   P: Do not allow employees to work if ill; protect foods from customers' coughs and sneezes.

4. Perfringens or CP

   O: Clostridium Perfringens

   F: Cells and toxins

   S: Reheated meats, sauces, stews, casseroles

   P: Keep cooked foods at an internal temperature of 140 degrees F or higher; reheat leftovers to internal temperature of 165 degrees F or higher.

5. Norwalk Virus

   O: Virus

   S: Spread almost entirely by poor personal hygiene of food service employees. It is found in human feces, contaminated water or vegetables fertilized with manure.

   P: The virus can be destroyed by high cooking temperatures, but not by sanitizing solutions or freezing.

6. Salmonella

   O: Salmonella

   F: Cells

   S: Poultry, eggs, milk, meats, fecal contamination

   P: Thoroughly cook all meat, poultry, fish and eggs; avoid cross-contamination with raw foods, maintain good personal hygiene.

7. E. coli or 0157

   O: Escherichia coli 0157

   F: Cells and toxins

   S: Any food, especially raw milk, raw vegetables, raw or rare beef, humans

   P: Thoroughly cook or reheat items.

8. Trichinosis

    O: Parasitic worms

    S: Eating undercooked game or pork infected with trichina larvae.

    P: Cook foods to a minimum internal temperature of 137 degrees F for 10 seconds.

9. Anisakiasis

    O: Parasitic roundworms

    S: The organs of fish, especially bottom feeders or those taken from contaminated waters. Raw or undercooked fish are often implicated.

    P: Fish should be thoroughly cleaned immediately after being caught so that the parasites do not have the opportunity to spread. Thoroughly cook to a minimum internal temperature of 140 degrees F.

10. Listeriosis

    O: Listeria Monocytogenes

    F: Cells

    S: Milk products, humans

    P: Avoid raw milk and cheese made from unpasteurized milk.

11. Staphylococcus

    O: Staphylococcus Aureus

    F: Toxins

    S: Starchy foods, cold meats, bakery items, custards, milk products, humans with infected wounds or sores

    P: Wash hands and utensils before use; exclude unhealthy food handlers; avoid having foods at room temperature.

2F. Matching

1. c   7. h

2. e   8. g

3. a   9. h

4. f   10. c

5. a   11. d

6. a   12. b

➤ CHAPTER 3 NUTRITION

3A. Terminology

Answers will not be provided in answer key. All answers can be found in text.

3B. The Chef's Role in Nutrition

1. a. Use proper purchasing and storage techniques in order to preserve nutrients.

    b. Offer a variety of foods from each tier of the food pyramid so that customers have a choice.

    c. Offer entrees that emphasize plant instead of animal foods.

    d. Offer dishes that are considerate of special dietary needs such as low fat or low salt.

    e. Use cooking procedures that preserve rather than destroy nutrients.

2. a. Ordering appetizers instead of entrees to control quantity of food

    b. Requesting half orders or splitting a full order

    c. Asking that dressings and sauces are served on the side

    d. Asking that a different cooking method be used

3. a. Reduce the amounts of the ingredient(s).

    b. Replace the ingredient(s) with a substitute that will do the least to change the flavor and appearance of the dish.

    c. Eliminate the ingredient(s).

3C. Roles of Nutrients on Health

1.	e	5.	b
2.	g	6.	a
3.	c	7.	d
4.	f		

3D. Essential Nutrients

1.	d	6.	c
2.	b	7.	a
3.	a	8.	b
4.	c	9.	b
5.	d	10.	d

3E. Parts of a Food Label

a. Serving size

b. Calories from fat

c. Daily values

d. Percentage of Daily values

e. Calories from fat

(See text for the description of each section of the food label)

3F. The Food Guide Pyramid

1.	e	5.	a
2.	h	6.	d
3.	f	7.	h
4.	c	8.	b

3G. Chapter Review

1. True

2. False   (p. 40)   Age, gender and health determine the various nutrients and the different amounts our bodies need.

3. True

4. False   (p. 50)   RDA stands for *Recommended Dietary Allowances*.

5. True

6. True

7. False   (p. 56)   A chef should be able to prepare and serve food that meets the high standards for health demanded by some patrons, while maintaining the flavor and appearance important to everyone.

8. False   (p. 42)   Dietary cholesterol is found only in foods of animal origin.

9. True

10. True

11. False   (p. 53)   Any artificial sweetener *cannot* be substituted for sugar when preparing baked goods. For instance, aspartame breaks down when heated and therefore cannot be used to sweeten foods that will be cooked.

12. True

13. False   (p. 42)   Dietary fibers are considered complex carbohydrates.

14. True

15. True

16 False   (p. 42)   While people are usually instructed to lower their cholesterol levels when they become too high because they are a risk factor for heart disease, a certain amount of cholesterol in the diet can actually be good. Because of the role cholesterol plays, it is a regulatory substance that aids in bodily functioning.

17. False   (p. 42, 46)   Vitamins and minerals provide no caloric value to the body and yet are considered essential nutrients because they are important to the body in generating energy from the foods we eat.

18. False   (p. 42)   While butter is solid at room temperature, it has not been hydrogenated to get to that state; it is a fat that is largely made up of saturated fatty acids, thus resulting in its solid characteristics.

3H. Putting It All Together

1. Answers will vary. See pages 49-51 in this chapter in order to determine the accuracy of your responses to this question.

2. Answers will vary based on the individual, but may include some of the following and more: ingredients naturally high in fat such as avocados or dark chocolate may be healthful if eaten in moderation. Other foods may be healthy until we add fat through the cooking process: such as frying foods, cooking eggs or meats. Others may include alcoholic beverages such as beer, which may increase bone density, or red wine, which may reduce the incidence of heart disease. The list can be extensive and might be shared as a discussion within your class based on each person's personal food preferences and eating habits.

➢ CHAPTER 4 ANSWER KEY

4A. Terminology

Answers will not be provided in answer key. All answers can be found in text.

4B. Units of Measure

1. 16 oz.
2. 28.35 g.
3. 435.6 g. = .4536 kg.
4. 1,000 g.
5. 035 oz.
6. 35 oz. = 2.187 lb./2 lb.
7. 16 tbsp. = 8 fl oz.
8. 1 qt. = 32 fl oz.
9. 1/2 gal. = 4 pt.
10. 32 tbsp. = 16 fl oz.
11. 8 tbsp. = 24 tsp.
12. 1 pt. = 1/2 qt.
13. 1/4 c. = 4 tbsp. = 12 tsp.
14. 128 fl oz. = 4 qt. = 16 c.
15. 3 pt. = .38 gal.

4C. Conversion Factors

1. 2.5
2. .7
3. .58
4. 2.5
5. .25
6. .75
7. a. .6; b. 120
8. 10

4D. Conversion Problems

1. Larger quantities may require the use of a mixer versus hand mixing. Also the mixing time may need to be modified accordingly.
2. The difference in surface area between using a saucepan versus a tilting skillet impacts evaporation. Thickness of the liquid should be modified accordingly.
3. Some recipes may have errors. Read recipes carefully and draw on professional knowledge to compensate for such errors.
4. Cooking time of individual items should not vary according to volume. Cooking time will be affected by the evaporation rate due to equipment changes. Rely on professional knowledge to compensate for such changes.

4E. Recipe Conversion

	Conversion Factor Yield I .88	Yield I 28 portions 6 oz. Each 168 oz.	Conversion Factor Yield II 1.31	Yield II 84 portions 3 oz. each 252 oz.
Butter		3.08 oz.		4.59 oz.
Onion		10.56 oz.		15.72 oz.
Celery		2.2 oz.		3.28 oz.
Broccoli		42.24 oz. = 2.64 lb.		62.88 oz. = 3.93 lb.
Chicken velouté		128 fl oz. = 3.52 qt.		5.24 qt.
Chicken stock		64 fl oz. = 1.76 qt.		2.62 qt.
Cream		21.12 fl oz.		31.44 fl oz.
Broccoli Florets		7.04 oz.		10.48 oz.

4F. Unit Costs

1. 53 cents
2. 25 cents
3. $3.25
4. 15 cents
5. 4 cents
6. 58 cents
7. $1.97
8. $4.02

4G. Cost Per Portion

1. $4.50
2. $4.13
3. 25 cents
4. 75 cents
5. a. $2.25
    b. $1.13
6. a. $4.07
    b. 140 oz.
    c. 28
    d. $1.27
7. a. 64 cents
    b. 112
8. $2.16

4H. Controlling Food Costs

1. The menu should be designed based upon customer desires, space, equipment, ingredient availability, cost of goods sold, employee skills and competition. Include all personnel when planning the menu.
2. Correct purchasing techniques help to control cost. Purchase specifications and periodic quotes help to insure value for money.

3. The person signing for the goods should be the person who actually checked them. Freshness, quality and quantity should always be checked.

4. Proper storage is crucial to prevent spoilage, pilferage and waste. Use the FIFO method.

5. Maintaining ongoing inventory record sheets helps the ordering process and proper stock rotation.

6. Standardized portions are a key factor in controlling food cost. Once an acceptable portion size has been decided, it must be adhered to.

7. An accurate sales history helps to prevent over production, but the chef must also be able to use leftovers to lower food cost.

8. Front of the house personnel must be trained to avoid loss of sales due to free meals or spilled foods.

4I. Chapter Review

1. True

2. False   (p. 57)   Portion or balance scales are commonly used in commercial kitchens to determine weights of ingredients. They are the most accurate form of measurement.

3. True

4. True

5. False   (p. 54)   For a meal served in the European tradition, the salad would be presented as a palate cleanser after the main dish and before the dessert.

6. True

7. False   (p.72)   Figuring the portion of everything on the plate is just one of may things that may be considered to determine the selling price.

8. False   (p.72)   The minimum selling price should $8.00

4J. Putting it All Together

1. For guests experiencing a tasting menu is an opportunity for a new and unique gastronomic journey when a chef carefully produces a special menu around an ingredient in season or another special occasion. The result is usually one of heightened sensory satisfaction and can help to strengthen the bond of customer loyalty because the experience is new and different. The more connected the guest becomes to an establishment in varied ways, through its service and food quality, the more committed that guest is for the long run; an important goal for any successful establishment.

   From the chef's perspective a tasting menu is the opportunity to create a special gastronomic experience that stretches beyond the establishment's menu, regardless of its format. The result may bring in additional business or it may simply be an added value for regular guests to enjoy. Creating a tasting menu may also be another opportunity to collaborate with purveyors, other chefs, and the sommelier to prepare a balanced menu that builds in experience and complexity from the first course to the last. Providing an additional challenge to the chef, it can help to keep the individual's creative juices flowing, thus increasing job satisfaction.

2. When an establishment invests the time to create standardized recipes, the resulting formulas are usually used to develop the ordering list for purveyors. In addition, the formulas are usually used so the chef can accurately determine overall food cost even before the actual sale is made. As a result if the chef does not follow accurate measuring techniques as per the

standardized recipe, it will be nearly impossible for him/her to determine accurate ordering quantities for the ingredients needed but could also result in running short of some of those ingredients needed for other menu items in the restaurant. Finally, taking away the predictability of food usage could also lead to an increased cost for producing a particular menu item.

In terms of the question regarding portion control, if a recipe has been standardized for the purpose of creating consistency in preparation from cook to cook as well as maintaining a consistent food cost in the establishment, then deviating from the predetermined portion guidelines could cause the restaurant to run out of the prepped item prematurely. From the guests' perspective such a practice could lead to varying perceived value; when inaccurate portioning techniques are followed the guest will feel s/he got a great value if the portion size is huge but only days later could feel cheated if the portion served is unusually small when ordering the same menu item.

3. If a chef is unable to accurately convert recipes the result could be an increase in production cost because s/he is not following the predetermined measures, waste of food by producing more portions than needed, not having enough of the menu item needed (a particular problem if serving a large party on a fixed menu), or simply the recipe/formula not producing the desired quality in the final menu item. To illustrate the final example, this is especially a problem in baking, where if the quantity of an ingredient such as shortening is inaccurately calculated, the resulting finished product fails to resemble the same quality and sensory appeal as the original recipe. Accurately converting recipes and carefully employing proper measuring techniques should result in the final quality of a menu item being identical to the original recipe, whether it is produced for 4 or 400 people.

➤ CHAPTER 5 ANSWER KEY

5A. Terminology

Answers will not be provided in answer key. All answers can be found in text.

5B. Equipment Identification

1. Zester

   Removing zest from citrus fruits

2. Straight spatula

   Applying frosting to cakes

3. Grill spatula

   Lifting hot food items from pan or grill

4. Meat mallet

   Flattening or tenderizing meat

5. Chef's fork

   Serving meats

6. French or chef's knife

   All purpose chopping

7. Rigid boning knife

   Separating meat from bone

8. Paring knife

   Detailed cutting of curved surfaces, namely vegetables

9. Flexible slicer

   Slicing meats and fish

10. Butcher knife/scimitar

    Fabricating raw meat

11. Steel

    Honing blade between sharpenings

12. Stock pot with spigot

    Making large quantities of soup or stock

13. Rondeau/brazier

    Stove top cooking for large amounts of food

14. Sautoir

    Stove top cooking for small amounts of food

15. Sauteuse

    Stove top cooking for sautéing

16. Wok

    Stir-frying and sautéing

17. Full hotel pan

    Holding food during service, baking, roasting, poaching

18. Drum sieve

    Sifting flour or straining

19. China cap

    Straining liquids

20. Skimmer

    Skimming stocks

21. Spider

    Removing particles from hot fat

22. Food mill

    Pureeing and straining foods

23. Mandolin

    Slicing small quantities of vegetables

24. Heavy duty blender

    Preparing smooth drinks and purees, chopping ice

25. Stack oven

    Baking

26. Tilt skillet

Large cooking utensil that can be used as a stockpot, brazier, fry pan, griddle or steam table

27. Steam kettle

Making stocks, soups, custards or stocks

28. Deep fat fryer

Deep frying foods

29. Insulated carrier

Keeping food hot during transportation

5C. Short Answer

1. a. Easily cleaned
   b. Nontoxic food surfaces
   c. Smooth surfaces
   d. Smooth and sealed internal surfaces
   e. Nontoxic coating surfaces
   f. Easily cleaned
2. a. Is it necessary for production?
   b. Will it do the job in the space available?
   c. Is it the most economical for the establishment?
   d. Is it easy to clean and repair?
3. a. Carbon steel
   b. Stainless steel
   c. High carbon stainless steel

5D. Matching

1. I   6. d
2. h   7. c
3. a   8. f
4. e   9. j
5. g   10. k

5E. Fill in the Blank

1. Utility   4. Buffalo chopper
2. Tang   5. Wooden
3. Griddle   6. Scimitar

5F. Calibrating a Stem-Type Thermometer

1. Fill a glass with shaved ice, then add water.

2. Place thermometer in the slush and wait until the temperature reading stabilizes. Following manufacturer's directions, adjust the thermometer's calibration nut until the temperature reads 32° F.

3. Check the calibration by returning the thermometer to the slush.

4. Repeat the procedure by substituting boiling water for the slush and calibrate the thermometer at 212° F.

5G. Chapter Review

1. False	(p. 84)	Stem-type thermometers should be calibrated after dropping.
2. False	(p. 96)	Ventilation hoods should be cleaned by professionals.
3. True		
4. False	(p. 96)	Class A fire extinguishers are used for fires caused by wood, paper, cloth or plastic.
5. True		
6. False	(p. 93)	Because a steam kettle's sides are heated as well as the bottom, it heats food more quickly than a pot sitting on a stove.
7. True		
8. False	(p. 79)	Seamless plastic or rubber parts on food service equipment are important from a standpoint of sanitation.
9. False	(p. 96)	A Class K fire extinguisher is also able to handle grease, fat and cooking oil in commercial cooking equipment as well as class B.
10. False	(p. 85)	Silicone bakeware is extremely versatile because one can bake in it as well as produce candy and chocolate products. In addition to having the capability of maintaining integrity up to 485° F, it can also be frozen.
11. True		
12. False	(p. 93)	Induction burners are gradually gaining popularity, as cook surfaces on buffet lines (omelet stations for example), but also in bakeshops where excessive heat that results from the cooking process is undesirable due to products like chocolate that cannot handle extreme fluctuations in temperature.
13. False	(p. 97)	Kitchen designers must also pay attention to the volume of food the operation intends to produce.

5H. Putting It Together

1. Alexis Soyer is considered by many as the father of the modern celebrity chef because he was a flamboyant, talented and egocentric showman. While he was a renowned chef in terms of the foods he prepared, he was also known for his kitchen designs that were ahead of their time. The designs helped to create a more healthy environment for the employees to work in while also increasing the function and flow for service. Part of his design included using the most modern cooking equipment available and linking the classic brigade system employed in a particular kitchen to the flow and design in order to maximize efficiency.

2. The food should be fully heated to an internal temperature of 165° F before being placed in the carrier. How long the food will be stored in the carrier will determine whether a heat source is needed to maintain the internal temperature of the food. One should take care to regularly check the internal temperature of the food product to ensure that it is kept above the temperature danger zone prior to serving.

3. Food should be stored and handled in a completely different area than any type of chemical. As a result, some clear cut safety measures you can take include never storing cleaning supplies or other chemicals with or near foods; never storing chemicals in a container that originally held food and vice versa and finally, keep chemicals and cleaners in properly labeled containers.

➤ CHAPTER 6 ANSWER KEY

6A. Terminology

Answers will not be provided in answer key. All answers can be found in text.

6B. Knife Safety

1. Always use the correct knife for the job.
2. Always cut away from yourself.
3. Always use a cutting board—never glass, marble or metal.
4. Keep knives sharp—a dull knife is more dangerous than a sharp one.
5. Do not attempt to catch a falling knife—step back and allow it to fall.
6. Never leave a knife in a sink of water—anyone reaching into the sink may get cut.

6C. Cuts of Vegetables

1. 1/8 inch x 1/8 inch x 1 to 2 inches
2. 1/4 inch x 1/4 inch x 2 inches
3. 3/8 inch x 1/2 inch x 1/2 inch
4. 1/8 inch x 1/8 inch x 1/8 inch dice
5. 1/4 inch x 1/4 inch x 1/4 inch dice
6. 3/8 inch x 3/8 inch x 3/8 inch dice

Similarities: 1. Brunoise comes from a julienne 2. Small dice comes from a batonnet

6D.   Fill in the Blank

1. Two
    Knife tip
    Wrist
2. Tip
    Rocking
3. Away
    Steel
    Glass
    Marble

4. Heel
    Coarsest
    Finest
5. Root

6E. Dicing an Onion

1. Remove the stem end with a paring knife, keeping the root intact. Peel off the outer skin without wasting too much.

2. Cut the onion in half through the stem and root.

3. Cut thin lines from the root towards the stem end, without cutting through the root.

4. Make as many cuts as possible through the width of the onion without cutting through the root.

5. Cut slices perpendicular to the other slices, producing diced onion.

6F. Chapter Review

1.	False (p. 104)	One must exert more effort using a dull knife, therefore increasing the chances of injury.
2.	True	
3.	False (p. 105)	A steel is generally used to straighten the edge of a knife.
4.	False (p. 111)	An allumette possesses different dimensions which more resemble a julienne, only out of potato.
5.	False (p. 105)	Only water should be used to moisten the stone.
6.	True	
7.	True	
8.	True	

6G. Putting It All Together

1. It takes much practice to perfect knife skills to a point where the resulting foods are both beautiful and produced in a time efficient manner. However, once those skills are mastered they can be a true way of separating your foods from that of your competitor. Combined with a skill for purchasing only the best ingredients, properly handling them, applying cooking methods in a masterful way, and presenting the final dish with great attention to detail, executing excellent knife skills shows a chef's desire to produce only the very best foods for the clientele.

2. Hand washing, rinsing and sanitizing knives after contact with *each* food is vital in order to prevent cross-contamination. The very definition of cross-contamination, transmitting characteristics including bacteria from one food to another via contact by hands, knives and cutting board, requires careful monitoring of the washing and sanitation process in order to prevent it from occurring.

➢ CHAPTER 7 ANSWER KEY

7A. Terminology

Answers will not be provided in answer key. All answers can be found in text.

7B. Discovering Tastes

1. d
2. h
3. g
4. a

5. f
6. c
7. b
8. i

7C. Categorizing Flavorings

Herb
1. Cilantro
2. Oregano
3. Lavender
4. Thyme
5. Lemon grass

Spice
6. Paprika
7. Coriander
8. Ground mustard
9. Capers
10. Black pepper
11. Garlic

7D. Herbs and Spices

1. a
2. c
3. b
4. d
5. d
6. a
7. c
8. b

9. d
10. a
11. b
12. c
13. d
14. c
15. b

7E. Short Answer

1. Japan
2. a. Taste buds
   b. Back of throat
   c. Roof of mouth
3. Saliva
4. a. At the top of the nasal cavity (the area within our head that collects the air that has been brought in through our nostrils)
   b. The area at the back of our throats contains receptors that take the aromas a 'back way' up through the nasal cavity to the olfactory receptor. This greatly enhances our ability to observe the taste of a food or beverage once it has been placed in one's mouth.

5. a. Season hot foods when they are hot and cold foods when they are cold because a food's temperature can affect how it tastes; foods served at warm temperatures offer the strongest tastes.

   b. A food with a thicker consistency will take longer to reach its peak intensity and will appear to have less flavor.

   c. A basic taste can be balanced or enhanced by using a small quantity of an opposing taste added to it. For example, lemonade that is too sour can be balanced by adding a small amount of sweetener. In the process the sour taste seems less extreme, is more palatable, and the sweetness adds a welcome taste as long as the lemonade hasn't been oversweetened.

   d. A moderate amount of fat in a food can help to moderate the release of flavor compounds as the food is chewed and mixed with saliva; too little fat and the flavors may disappear too quickly, too much fat and it may inhibit the sense organs' ability to absorb the flavor compounds to the fullest of their ability.

   e. Another sense organ humans possess: the ability to see can play a tremendous role in our enjoyment of food even before we've taken a bite. A mere glimpse of a food can tell us a lot about color, texture, portion size and can help us to identify ingredients that may be appealing to us. As color level of a food increases to match normal expectations, our perception of taste and flavor intensity increases as well.

6. While each work scenario poses its own challenges, working in a nursing home may be a welcome change for some chefs. Doctor prescribed diets create a special challenge for the chef, who often works hand-in-hand with a Dietician, to create nutritious, satisifying meals to maximize the changing nutritional needs of aging clients. The mere fact that most of the clients are aging and therefore have a decreased sensitivity to taste and smell cause the chef to strategize ways to flavor foods that will create interest in the foods while still meeting dietary requirements.

7. See page 146 (Still and sparkling wines)

8. In an attempt to reduce the amount of hard liquor consumed in the United States, Thomas Jefferson, quite a French wine enthusiast, encouraged Americans to plant European wine grapes (Vitis Vinifera) in an effort to increase wine production and offset the alcoholic beverages being consumed. As Vitis Vinifera grape vine cuttings were brought back and forth between Europe and America, a very destructive vine louse from America called phylloxera took hold in Europe. By the late 1800s many of the Vitis Vinifera vineyards in France and elsewhere were destroyed and numerous French Vintners left their country for a career elsewhere.

9. The basic guidelines are listed below, descriptions are found on page 153.

   a. Match tastes

   b. Match strengths

   c. Match opposites

   d. Match origins

10.

Alcoholic Beverage	Production Method	Base Ingredient	Additional Ingredients
a. German Beer	Fermentation	Water, hops, malted barley and yeast	By law not allowed, but in U.S.A. and elsewhere other ingredients may be used.
b. Cognac	Fermented grape juice and/or pulp and/or skin that later goes through a double distillation process	Grape Brandy—often a blend of a variety of wines from different vintners and years	Generally no additional ingredients, however it is often aged 2.5 to 6 plus years.
c. Gin	Liquid made from grains, vegetables or the like—is fermented, then the water component is cooked off and the alcohol is concentrated through distillation.	Grains	Main flavoring is juniper berries but may also contain herbs, peels and spices.
d. Rum	Liquid made from grains, vegetables or the like—is fermented, then the water component is cooked off and the alcohol is concentrated through distillation.	Sugar cane	Occasionally rums are flavored, such as with spices, and they may also vary in color from light to gold to dark.
e. Vodka	Liquid made from grains, vegetables or the like—is fermented, then the water component is cooked off and the alcohol is concentrated through distillation.	Potatoes, fruits, grains and other plant products may be used but most of the world's vodka is made from wheat.	Recently an increasing number of vodkas are flavored —either in the mash during distillation or adding flavors afterwards.

Alcoholic Beverage	Production Method	Base Ingredient	Additional Ingredients
f. Tequila	Liquid made from grains, vegetables or the like—is fermented, then the water component is cooked off and the alcohol is concentrated through distillation.	Agave sap	Generally none
g. Whiskey	Liquid made from grains, vegetables or the like—is fermented, then the water component is cooked off and the alcohol is concentrated through distillation.	Various grains that have been pounded and cooked into a mash. Type of grains used vary by whiskey type.	Additional ingredients may be added depending on the type of whiskey, as with a blended whiskey, where neutral spirits are added to the whiskey before aging. Also, there are many different types of whiskey, like blended, bourbon, Canadian, Irish, Rye and Scotch; ingredients added will depend on what type.
h. Liqueurs	Traditionally made from herbs, fruits, nuts, spices, flowers or other flavorings infused in a distilled alcohol base.	Neutral spirits such as brandy, rum or whiskey with varying degrees of added sugar	Flavors can vary widely and may be produced from infusion of natural ingredients or artificial flavorings.

7F. Chapter Review

1. True

2. True

3. False   (p. 131)   Use less dried herbs than you would fresh herbs in a recipe. Loss of moisture in the dried herbs strengthens and concentrates the flavors.

4. True

5. True

6. True

7. False   (p. 157)   Distilled vinegar is made from grain alcohol.

8. False   (p. 139)   Salt can be tasted easily but not smelled.

9. True

10. True

11.	False	(p. 131)	Spices have been used therapeutically, cosmetically, medicinally, ritualistically, as well as for culinary purposes.
12.	False	(p. 122)	This is a myth created by the misinterpretation of a German article written in the 1800s. You can in fact taste all taste compounds everywhere on your tongue.
13.	True		
14.	True		
15.	False	(p. 138)	Although limited in flavoring application, basil and dill work wonderfully as toppings to savory breads and bagels.
16.	True		
17.	False	(p. 145)	People have been consuming wines for thousands of years.
18.	False	(p. 146)	Chianti is not considered a noble grape varietal.
19.	True		
20.	True		
21.	False	(p. 148)	They may have a certain rich, complex flavor with cherry and raspberry undertones, but they will not be identical beverages. Differences in the conditions under which the grapes were grown or the techniques the vintners used to make the wines will create noticeable differences between them.
22.	True		
23.	True		
24.	False	(p. 151)	When selecting wines to use as flavorings. Avoid using anything called cooking wine since they are often inferior in quality and have added salt and other flavorings. Do not cook with a wine that you would not drink.
25.	True		
26.	False	(p. 153)	Ales are most heavily consumed in Britain and Belgium, not lagers.

7G. Putting It All Together

1. The chef could try a variety of things. First, s/he could suggest that the guest first adjust to eating smaller portion sizes of meats and to choose meats that are lower in cholesterol. For example, red meats in general are much higher in cholesterol than white meats or some fish. Second, if the guest wants to make a more drastic change, the chef could recommend a vegetarian dish, which due to the lack of animal products, would be naturally low in cholesterol. To create a meaty texture and flavor s/he could use ingredients that are naturally high in amino acid glutamates and therefore possess an umami taste profile: mushrooms, highly reduced vegetable stock, soy sauce, tomatoes and perhaps a small amount of cheese.

2. Ice milk contains minute amounts of fat compared to ice cream. While the two may contain the exact same amounts of sugar and flavorings, the lower fat content of the ice milk can cause it to dissolve more quickly in the mouth and for the flavorings to disappear from the palate more quickly than the ice cream. In addition, the fat in the ice cream helps to create a mouthfeel, or creaminess, which is naturally more appealing than the ice milk, coating the interior surfaces of the mouth, mixing with the saliva, and coating the sense organs more completely and for a longer period of time than the ice milk. Of course, if the ice cream is too rich with fat, it can actually be a negative characteristic since it can almost clog the sense organ's ability to taste all the elements of the dessert.

➤ ## CHAPTER 8 ANSWER KEY

8A. Terminology

Answers will not be provided in answer key. All answers can be found in text.

8B. Comparing Creams

1. e
2. d
3. b
4. a

8C. Cheese Identification

1. e	6. a	11. d	16. j
2. g	7. s	12. c	17. i
3. b	8. l	13. p	
4. m	9. r	14. n	
5. f	10. q	15. h	

8D. Milk Products

1. c	9. a
2. a	10. b, c
3. a	11. d
4. c	12. c
5. d	13. c
6. b	14. d
7. b	15. c
8. c	

8E. Chapter Review

1. True
2. False   (p. 170)   Margarine is not made from animal products and therefore does not contain cholesterol.
3. True
4. False   (p. 166)   By law, all Grade A milk must be pasteurized prior to retail sale.
5. True
6. True
7. True
8. False   (p. 168)   Yogurt is only as healthful, or low in fat, as the milk it is made from.

9. False	(p. 170)	Margarine is not a dairy product and is included in this chapter only because it is so commonly used as a substitute for butter. It is actually made from animal or vegetable fats or a combination thereof. Flavorings, colorings, emulsifiers, preservatives and vitamins are added before it is hydrogenated.
10. False	(p. 179)	Processed cheese food contains less natural cheese and more moisture than regular processed cheese. Often vegetable oil and milk solids are added to make the cheese food soft and spreadable.
11. True		
12. True		
13. True		
14. False	(p. 167)	Milk, as well as all perishable foods, should be refrigerated at 41° F or below.
15. True		
16. True		
17. False	(p. 171)	The FDA does allow the manufacture and distribution of rawmilk cheeses in the United States as long as they are aged more than 60 days at a temperature not less than 35° F.

8F. Putting It All Together

1. European-style whole butter would provide a creamier texture as a final liaison to the Poulet Sauce due to its higher fat content. As learned in Chapter Three, butter contains fat and therefore flavor and mouthfeel. Therefore, the larger amount of fat found in the European-style butter compared to traditional American-style butter would result in a richer, creamier sauce.

2. Since cholesterol is a lipid only found in animal products such as milks and creams, when fat is removed from a dairy product to create a lower fat version the new version becomes naturally lower in cholesterol as well as fat.

➢ CHAPTER 9 ANSWER KEY

9A. Terminology

Answers will not be provided in answer key. All answers can be found in text.

9B. Short Answer

1. Use one hand for dipping the food into the liquid ingredients and the other hand to dip into the dry ingredients.

2. a. Prepare the batter
    b. Pat the food dry and dredge in flour if desired
    c. Dip the item in the batter and place directly in the hot fat

3. a. Fat: $2.15
    b. Water: $0.43
    c. Milk solids: $0.11

4. a. Work patterns
   b. Tool/equipment needs
5. a. Assemble the mise en place.
   b. With left hand, place the food to be breaded in the flour and coat it evenly. With the same hand, remove the floured item, shake off the excess flour and place the floured item in the egg wash.
   c. With right hand, remove the item from the egg wash and place it in the bread crumbs or meal.
   d. With left hand, cover the item with crumbs or meal and press lightly to make sure the item is completely and evenly coated. Shake off the excess crumbs or meal and place the breaded item in the empty pan for finished product.
6. a. Fat: 12.8 oz.
   b. Water: 2.56 oz.
   c. Milk solids: 0.11 oz.

9C. Multiple Choice

1.	b	6.	c
2.	a	7.	c
3.	d	8.	a
4.	a	9.	b
5.	d	10.	a

9D. Chapter Review

1. True
2. False   (p. 191)   Beer batters contain beer for leavening as well as for flavor.
3. False   (p. 184)   Front of house personnel, such as waiters, must prepare mise en place that may include cutting drink garnishes, folding napkins, polishing flatware/silverware, and refilling salt and pepper shakers, to name a few.
4. False   (pp. 185-186)   A food's weight does not always equal its measurement in volume.
5. False   (p. 186)   Ghee contains the clarified fat and milk solids together and the milk solids are allowed to brown to enhance flavor. Milk solids are removed when making classical clarified butter.
6. True
7. True
8. False   (p. 190)   Breaded foods are usually cooked by deep fat frying or pan frying.
9. True
10. False   (p. 187)   The bread was likely too fresh as opposed to being stale. You should have used 2-4 day-old bread instead of fresh bread to make the fresh bread crumbs.
11. True
12. True

9E. Putting It All Together

1. A tamis is a piece of equipment that is comprised of a metal ring that has mesh on one side of the opening. Food, such as bread crumbs are placed in the tamis and by moving the equipment back and forth in a shaking motion, the bread crumbs are sifted, resulting in a fine and more consistently shaped and sized bread crumb.

2. Raw meats of any kind should be stored on the lower shelves in the refrigerator, below and raw produce, dairy, and other ingredients that may be considered ready-to-eat.

➢ CHAPTER 10 ANSWER KEY

10A. Terminology

Answers will not be provided in answer key. All answers can be found in text.

10B. Cooking Methods

Cooking method	Medium	Equipment
ex: Sautéing	Fat	Stove
1. Stewing	Fat then liquid	Stove (and oven), tilt Skillet
2. Deep-frying	Fat	Deep fryer
3. Broiling	Air	Broiler, salamander, rotisserie
4. Poaching	Water or other liquid	Stove, oven, steam jacketed kettle, tilt skillet
5. Grilling	Air	Grill
6. Simmering	Water or other liquid	Stove, steam jacketed Kettle, tilt skillet
7. Baking	Air	Oven
8. Roasting	Air	Oven
9. Steaming	Steam	Stove, convection steamer
10. Braising	Fat then liquid	Stove (and oven), tilt skillet

10C. Smoking Points

1. f
2. i
3. a
4. h
5. b
6. g
7. c
8. d

10D. Multiple Choice

1.	a	7.	d
2.	d	8.	c
3.	c	9.	a
4.	a	10.	b and d
5.	c	11.	a
6.	b	12.	c

10E. Short Answer

1. **Braising**   **Stewing**

    a. Large pieces of food   a. Smaller pieces of food

    b. Brown then simmer/steam   b. Brown or blanch then simmer/steam

    c. Cooking liquid covers 1/3-1/2   c. Cooking liquid completely covers

    d. Cooking time is longer   d. Cooking time is shorter

2. a. The food product must be placed in a basket or on a rack to allow for circulation of the steam.

    b. A lid should cover the steaming unit to trap steam and allow heat to build up.

3. a. Cut, pound or otherwise prepare the chicken breast.

    b. Heat a sauté pan and add enough fat to just cover the pan's bottom.

    c. Add the chicken to the sauté pan in a single layer, presentation side down first. Do not crowd the pan.

    d. Adjust temperature as needed to control browning, flip when half cooked.

    e. Turn the chicken breast.

    f. Cook until done.

4. a. Cut, trim or otherwise prepare the food to be poached.

    b. Bring an adequate amount of cooking liquid to the desired starting temperature. Place the food in the liquid.

    c. For submersion poaching, the liquid should completely cover the food.

    d. For shallow poaching, the liquid should come approximately halfway up the side of the food. Cover the pan with a piece of buttered parchment paper or a lid.

    e. Maintaining the proper temperature, poach the food to the desired doneness in the oven or on the stove top. Doneness is generally determined by timing, internal temperature, or tenderness.

    f. Remove the food and hold it for service in a portion of the cooking liquid or, using an ice bath, cool it in the cooking liquid.

    g. The cooking liquid can sometimes be used to prepare an accompanying sauce or reserved for use in other dishes.

10F. Matching

1. e
2. c
3. b

4. a
5. d
6. g

10G.   Chapter Review

1. False   (p. 198)   Heat is generated quickly and uniformly throughout the food. Microwave cooking does not brown foods, however, and often gives meats a dry, mushy texture, making microwave ovens an unacceptable replacement for traditional ovens.

2. True

3. False   (p. 201)   A wood-fired grill is an example of the radiation heat transfer method.

4. False   (p. 24)   Chapter 2: 165° F internal temperature

5. True

6. False   (p. 200)   In broiling the heat source comes from above the cooking surface.

7. False   (p. 205)   Deep-frying is an example of a dry heat cooking method.

8. True

9. False   (p. 204)   Stir-frying is a variation in technique to sautéing but it does not necessarily use any additional fat in the process.

10. False   (p. 209)   A court bouillon should be used when poaching or simmering foods.

11. True

12. True

13. True

14. False   (p. 24)   Chapter 2: 165° F internal temperature

15. False   (p. 206)   Battered foods that will stick to the wire frying baskets are usually cooked using the swimming method.

16. True

17. False   (p.206)   Filling the wire frying basket while it is hanging over the hot fat allows unnecessary salt and food particles to fall into the fat, shortening its life.

18. True

10H. Putting It All Together

1. Just as different metals are more efficient at conducting heat than others (for example, copper is better at conducting heat than stainless steel), foods cook at different rates in different media. So in the question above, water (liquid) is a more efficient cooking medium (and therefore a better conductor of heat) than air, a gas. However, when comparing these two media to metals, nearly all metals are superior conductors.

2. Chapter Three also explained that in addition to making a vegetable oil solid, hydrogenation reduces the tendency for rancidity (developing an off-flavor), therefore increasing shelf life. In the case of deep frying this concept is very important because the fat is used as the cooking medium and a certain percentage of it is absorbed by the foods that are fried in it. A chef may choose to fry foods in a hydrogenated frying oil because in the long run it may have a lesser chance of becoming rancid during the usable life of the fat, therefore preventing the worry that the fat (the cooking medium) will impart an off-flavor to the food. This is especially important when frying mildly flavored foods such as French fries.

3. The purpose of braising is to tenderize a tough cut of meat and therefore the first way to determine doneness should be to insert a braising fork into the center of the piece. If it is inserted and removed without resistance, it is tender and ready to serve, making it enjoyable to eat. The second determining factor is whether or not the meat has reached an internal temperature of 165° F, the upper end of the temperature danger zone, where if the meat is at this level of doneness it is also safe to consume.

➤ CHAPTER 11 ANSWER KEY

11A. Terminology

Answers will not be provided in answer key. All answers can be found in text.

11B. Stock Making Review

White Stock: reference page 222.
Brown Stock: reference pages 223-224.
Fish Stock: reference page 226.

11C. Mother Sauce Review

Mother Sauce	Thickener	Liquid
1. Béchamel	White roux	Milk
2. Velouté	Blond roux	White chicken/veal/fish stock
3. Espagnole	Brown roux	Beef stock
4. Tomato	Tomato purée	White stock
5. Hollandaise	Egg yolks (emulsion)	Butter

11D. Small Sauces

Mother Sauce	Ingredients Added
1. Béchamel	Scalded cream, lemon juice
2. Béchamel	Cheddar, Worcestershire sauce, dry mustard
3. Béchamel	Gruyere, Parmesan, scalded cream, butter
4. Béchamel	Heavy cream, crayfish butter, paprika, crayfish meat
5. Béchamel	Onion, sweated, cook and strain sauce
6. Chicken/veal velouté	Lemon juice and a liaison
7. Chicken velouté	Cream

Mother Sauce	Ingredients Added
8. Fish velouté	Diced shallots, dry white wine, butter and parsley
9. Fish velouté	Heavy cream, cayenne pepper, lobster butter, lobster coral
10. Fish velouté	Mushroom, liaison, strained
11. Chicken/veal velouté	Sauce Allemande, tomato paste, butter
12. Chicken/veal velouté	Sauce Allemande, heavy cream, mustard, horseradish
13. Chicken/veal velouté	Sauce Allemande, mushroom, shallots, cream, lemon juice, parsley
14. Chicken velouté	Sauce Supreme, glace de volaille, red pepper butter
15. Chicken velouté	Sauce Supreme, onion, butter, paprika, strain
16. Chicken velouté	Sauce Supreme, glace de volaille
17. Espagnole	Demi-glace, red wine, shallots, bay leaf, thyme, black pepper, butter, sliced poached beef marrow
18. Espagnole	Demi-glace, mushrooms, shallots, white wine, diced tomatoes, parsley
19. Espagnole	Demi-glace, white wine, shallots, lemon juice, cayenne pepper, tarragon
20. Espagnole	Poivrade sauce, (with bacon trimmings added to mirepoix), red wine, dash of cayenne
21. Espagnole	Demi-glace, madeira wine or ruby port
22. Espagnole	Demi-glace, red wine, shallots, strain
23. Espagnole	Demi-glace, truffles
24. Espagnole	Demi-glace, shallots, white wine, white wine vinegar, cornichons, tarragon, parsley, chervil
25. Espagnole	Demi-glace, mirepoix, bouquet garni, vinegar, white wine, crushed peppercorns, butter, strain
26. Espagnole	Demi-glace, onion, white wine, Dijon mustard, sugar, sliced pickles
27. Tomato sauce	Onion, celery, garlic, bay leaf, thyme, green pepper, hot pepper sauce
28. Tomato sauce	Mushroom, cooked ham, cooked tongue
29. Tomato sauce	Mushroom, onion, sliced black or green olives
30. Hollandaise	Shallots, tarragon, chervil, crushed peppercorns, whitewine vinegar, cayenne pepper. Garnish: tarragon
31. Hollandaise	Bearnaise sauce, tomato paste, heavy cream
32. Hollandaise	Bearnaise sauce, glace de viande
33. Hollandaise	Infused with saffron
34. Hollandaise	Orange juice, orange zest—blood oranges are traditional
35. Hollandaise	Heavy cream

11E. Short Answer

1. a. Start stock with *cold* water.
    b. *Simmer* the stock gently.
    c. *Skim* the stock frequently.
    d. *Strain* the stock carefully.
    e. *Cool* the stock quickly.
    f. *Store* the stock properly.
    g. *Degrease* the stock.
2. Hollandaise sauce: reference pages 235, 245.
3. a. Incorrect temperature of eggs and/or butter.
    b. Butter added too quickly.
    c. Egg yolks overcooked.
    d. Too much butter added.
    e. Sauce not whipped enough.
4. Thickening agents: reference pages 230-235.

11F. Matching

1. d, k, n
2. b, i, l
3. a, f, j
4. c, g, m
5. e, h, o

11G. Chapter Review

1. True
2. False   (p. 232)   To prevent lumps when making sauces add cold stock to hot roux or room-temperature roux to hot stock.
3. False   (p. 234)   Temperatures over 185° F will cause the yolks to curdle.
4. True
5. True
6. True
7. False   (p. 234)   Tempering gradually raises the temperature of a cold liquid, such as a liaison, by adding hot liquid.
8. True
9. True
10. False   (p. 225)   Fish stock should only simmer for 30 to 45 minutes.
11. False   (p. 231)   Antonin Carême is credited with developing the modern system for classifying hundreds of sauces.
12. True
13. True

14. False	(p. 228)	A nage is derived from a court bouillon.
15. False	(p. 221)	A china cap is the most appropriate piece of equipment.
16. True		
17. True		
18. False	(p. 219)	Beef bones are generally from an older animal and contain less collagen protein from cartilage and other connective tissue than veal bones or chicken bones, which in today's modern food service industry, tend to yield bones from a younger animal.
19. False	(p. 219, 227)	Vegetable stocks are not produced using bones as an ingredient and contain no collagen protein, thereby resulting in a thinner body than meat stocks.

11H. Putting It All Together

1. Since fat is lighter than liquid, it will naturally rise to the surface of the stock. If the fat is of animal origins, it will solidify into a mass that is impenetrable by air. Without ready access to air, one of the elements bacteria need to survive, it slows the growth of bacteria and may cause the stock to have a slightly longer shelf life provided the stock is stored at suggested temperatures.

2. Hot, freshly strained stock should be poured into clean, shallow metal containers and then chilled in an ice bath, stirring periodically during the cooling process. The stock should cool to 70° F (21° C) within two hours, then cool to 41° F (5° C) or below within four hours, cover and refrigerate; store cooked food above raw. Since the stock should be cooled in shallow containers (generally less than four inches deep) a cooling wand may not be an effective method to employ.

   Cooling wands, which may vary in size but generally tend to be two feet long at a minimum, are generally designed to be frozen and then inserted into a deep container of cooling foods, such as a five gallon storage bucket or larger. Because the depth of the cooling food is so great, it will take a longer time to cool in an ice bath than if it is in a metal container that is four inches deep or less. As a result, the cooling wand can expedite the cooling process by being inserted into the center of the deep container, therefore reaching the center of the cooling food product that may not otherwise have been penetrated as quickly by the sole use of an ice bath. (p. 29 chapter two)

➤ CHAPTER 12 ANSWER KEY

12A. Terminology

Answers will not be provided in answer key. All answers can be found in text.

12B. Short Answer

1. Broth preparation: reference pages 279-280.
2. Consommé ingredients and procedure: reference pages 282-285.
3. a. If consommé is allowed to boil or if it is stirred after the raft has been formed.
    b. Stock was not degreased.
    c. Poor quality stock.
    d. Onion brulée omitted.

4. a. Thoroughly chill and degrease the consommé.

   b. Lightly beat four egg whites per gallon of consommé and combine with the cold consommé.

   c. Slowly bring the consommé to a simmer, stirring occasionally. Stop stirring when the egg whites begin to coagulate.

   d. When the egg whites are completely coagulated, carefully strain the consommé.

5. Cream soup ingredients and preparation: reference pages 285-287.

6. a. Never add cold milk or cream to a hot soup.

   b. Add milk or cream just before service.

   c. Do not boil soup after cream has been added.

   d. Utilize a thickening agent in the soup; even small quantities will help stabilize

12C. Soup Review

1. Seven common categories of soup: reference page 278.

2. a. Beef broth and consommé both have the same base—beef stock; however, beef broth uses meat and vegetables to give it a fuller flavor and consommé uses clearmeat.

   b. A cream of mushroom soup uses a roux to thicken the soup, but lentil soup uses a purée of the vegetable to thicken the soup. They may both use stock to form the base for flavor and both may use cream to finish the soup.

   c. Both are cold soups, however gazpacho uses uncooked ingredients and the cold consommé uses ingredients that are cooked, strained and cooled before service.

12D. Chapter Review

1. True

2. True

3. True

4. False   (p. 282)   A consommé is a clarified broth.

5. False   (p. 285)   Cream soups are thickened with a roux or other starch.

6. True

7. False   (p. 292)   Cold soups should be served at 40° F/4° C or below.

8. False   (p. 285)   If the consommé is insufficiently clear, a clarification can be performed.

9. False   (p. 293)   The cold temperature at which the soup is served dulls the soup's flavor and therefore more seasoning is required.

10. True

11. True

12. False   (p. 291)   Not all chowders contain milk or cream.

13. True

14. False   (p. 34)   Thick soups do scorch easily, but the food safety challenge is properly cooling and reheating the thick products as quickly as possible; both processes take longer with the thicker consistency.

15. True

16. False		The conversion factor would be 0.375.
17. True		
18. True		
19. False	(p. 293)	Vichyssoise contains 3 PHF: potatoes, chicken stock and heavy cream.
20. True		

12E. Putting It All Together

1. From a storage of raw product perspective, a chef must be certain that the raw ingredients to be used in the production of the soup have been sourced from a reputable supplier and stored at the proper temperature. Once the product has arrived in house, it should be washed thoroughly to remove contaminants, soil and pesticides. The utmost care of handling the food product must be taken so as to prevent cross-contamination since the soup will never be cooked, therefore eliminating the opportunity to kill food-born pathogens. Finally, the chef must work quickly during the preparation and service periods in order to keep the ingredients out of the temperature danger zone as much as possible. The finished product must be stored and served below 41° F.

2. Conduct the final seasoning of hot soups when they are hot and cold soups when they are cold (and at the serving temperature so that seasoning is accurate). Since warm foods amplify flavors more readily, extra care must be taken to adjust the seasoning of a cold soup to make sure it is perfect.

3. The cost of ingredients to produce the artichoke soup are significantly higher than that of the cherry soup. For example, the more costly ingredients include the fresh artichokes, olive oil, leeks, shallots and truffles and while the ingredient cost for the opal basil oil and vegetable stock is not high, the labor and time needed to mise en place both of these items is fairly significant. The preparation time needed to put the soup together must also be noted as well as the cost of the heat to cook the soup.

   In contrast, the cherry soup contains fairly inexpensive ingredients compared to the artichoke soup. Aside for the champagne which could be fairly costly if the chef chooses a good quality wine to complement the delicate flavors of the fruit in the soup and the crème fraiche, the number of costly ingredients is significantly lower and the labor and time required to produce the soup is noticeably less.

➤ CHAPTER 13 ANSWER KEY

13A. Terminology

Answers will not be provided in answer key. All answers can be found in text.

13B. Fill in the Blank

1. Good quality
2. Low, long
3. Soft, very red; firm, non-red
4. Dredged (in) flour
5. Carryover, retain (more) juices

6. Barding, larding
7. Flavor, connective tissue
8. Insects, bacteria, parasites
9. Tough
10. Against

13C. Matching

1. d
2. e
3. f
4. b
5. c

13D. Cooking Methods

Reference pages 326-346.

13E. Chapter Review

1. False	(p. 324)	Fresh meats should be stored at 30-35° F.
2. False	(p. 322)	Green meats are meats that are frozen before rigor mortis has had an opportunity to dissipate.
3. True		
4. True		The USDA stamp only insures that the meat is processed in a sanitary way.
5. False	(p. 322)	USDA *Prime* cuts are used for the finest establishments.
6. False	(p. 322)	Yield grades apply only to lamb and beef.
7. True		
8. True		
9. False	(p. 324)	Meat will hold in a vacuum package for 3-4 weeks under refrigeration.
10. True		
11. False	(p. 322)	The marbling in the meat is the principal factor in determining meat quality.
12. True		
13. True		
14. True		
15. False	(p. 321)	USDA inspection only insures that the meat was handled in a sanitary fashion.
16. True		
17. False	(pp. 329-330)	Due to its vast size, a steamship round should be removed from the oven at approximately 125° F to take into account carryover cooking.
18. False		The conversion factor is 2.5.

19. False   (p. 325)   A whole stuffed chicken is not such a lean cut of meat that it needs lardoons of fat inserted into its center in order to add flavor and moisture. If anything, a whole stuffed chicken needs to be trussed in order to maintain its shape and even cooking.

20. True

13F. Putting It All Together

1. The time/temperature principle applies to all areas of food service, from the time an animal is slaughtered to the time the food is plated and presented to the guest. If a primary supplier does not follow HACCP procedures properly then a chef's job is far more challenging because s/he has an even smaller window of time to work with food to prepare it for the guest and still keep it safe for human consumption. Providing the primary supplier follows HACCP procedures properly, the chef should have a maximum of 4 hours to work with the food but the less time taken, the better.

2. The text does not recommend consuming these meats rare because at this temperature one cannot ensure that harmful bacteria have been killed and therefore will not harm the consumer. With past risks of trichinosis in pork in particular, you'll notice that one should cook pork to a minimum internal temperature of 160° F.

3. Au jus refers to serving a roasted piece of meat, poultry or game with the natural pan drippings that result from the cooking process. The juices may be strained before serving but are not thickened.

➤ CHAPTER 14 ANSWER KEY

14A. Terminology

Answers will not be provided in answer key. All answers can be found in text.

14B. Primal Cuts of Beef

1. Chuck
2. Rib
3. Short loin
4. Sirloin
5. Round
6. Flank
7. Short plate
8. Brisket and shank

14C. Cuts from the Round

Subprimal/Fabricated Cut	Cooking Process/Use
1. Inside (top) round	Roast
2. Eye of round	Braise
3. Outside (bottom) round	Braise
4. Knuckle or tip	Roast
5. Leg or round bone	Simmering--stocks, soups and consommé

14D. Cuts of Beef and Applied Cooking Methods

	Cooking Method	Subprimal/ Fabricated Cut	Primal Cut
1.	Combination (braise/stew)	Chuck	Chuck
2.	Combination (braise)	Shank	Brisket and shank
3.	Dry heat (broil/grill/roast/sauté)	Strip loin	Short loin
4.	Dry heat (broil/grill/roast)	Ground beef	Chuck
5.	Combination (braise)	Flank steak	Flank
6.	Moist heat (simmer)	Brisket	Brisket and shank
7.	Dry heat (broil/grill/roast)	Tenderloin	Short loin
8.	Dry heat (broil/grill)	Flank steak	Flank
9.	Dry heat (broil/grill/roast)	Tenderloin	Short loin
10.	Dry heat (roast)	Steamship/top round	Round
11.	Combination (braise)	Top round	Round
12.	Dry heat (broil/grill)	Skirt steak	Short plate
13.	Combination (stew)	Stew meat	Chuck
14.	Combination (braise/stew)	Ground beef	Chuck
15.	Dry heat (broil/grill/roast/sauté)	Strip loin	Short loin

14E. Multiple Choice

1. b
2. c
3. d
4. a
5. a
6. c
7. 2, 1, 3

14F. Matching I

1. c
2. e
3. a
4. b

14G. Matching II

a. 3
b. 1
c. 2
d. 2
e. 1
f. 3
g. 1
h. 2
i. 3

14H. Chapter Review

1. True

2. False   (p. 350)   The chuck has a high proportion of connective tissue which makes it more flavorful than the tenderloin.

3. True

4. False   (p. 352)   Prime rib refers to the fact that the rib is made up of the majority of the primal cut from which it comes.

5. True

6. True

7. True

8. False   (p. 352)   Pastrami is made by curing and peppering the brisket.

9. True

10. False   (p. 352)   Grinding the shanks is appropriate, but the meat is typically used in soups and stocks.

11. False   (p. 374)   Four pounds of meat are needed since you're not increasing the yield of the total recipe, only the size and number of portions that result.

12. True

14I. Putting It All Together

1. Trussing, barding

2. No, it would not make good financial sense to braise a strip loin. First of all, as a chef you will be paying good money for a strip loin because it is a tender piece of meat with good marbling and therefore will be in high demand on your menu. You will not want to perform such a long, drawn out cooking process on such a cut, partially because of what it costs you in time and money to braise and partially because you don't need to. In most restaurants a good quality steak is in high demand and in order to make the most of that entrée you want to be able to have minimal processing of that product so that your profit margin is greater. If you were determined to braise you would be better off spending less money on a piece of meat that would benefit from the braising process because the resulting product would be flavorful. Because your food cost was less on the later braised product, your profit margin will be greater than if you had used strip loin.

➤ CHAPTER 15 ANSWER KEY

15A. Terminology

Answers will not be provided in answer key. All answers can be found in text.

15B. Primal Cuts Identification

Answers will not be provided in answer key. All answers can be found in text pages 380-383.

15C. Primal Cuts of Veal

1. Shoulder
2. Rib
3. Loin
4. Leg
5. Foreshank and breast

15D. Cuts of Veal and Applied Cooking Methods

	Cooking Method	Subprimal/ Fabricated Cut	Primal Cut
1.	Combination (stew)	Cubed veal	Shoulder
2.	Dry heat (broil/grill/roast)	Rib eye	Rib
3.	Combination (braise)	Breast	Foreshank and breast
4.	Dry heat (broil/grill/sauté)	Loin chops	Loin
5.	Combination (stew)	Cubed veal	Shoulder
6.	Combination (braise)	Sweetbreads	Offal
7.	Combination (braise)	Kidneys	Offal
8.	Dry heat (broil/grill)	Ground veal	Shoulder
9.	Combination (braise)	Ground veal	Shoulder
10.	Dry heat (broil/grill/roast/sauté)	Veal tenderloin	Loin
11.	Dry heat (roast/sauté)	Top round	Leg
12.	Dry heat (broil/grill/sauté)	Calves' liver	Offal
13.	Combination (braise)	Hind/Foreshank	Leg/Foreshank and breast
14.	Dry heat (roast/sauté)	Leg	Leg
15.	Moist heat (simmer)	Hindshank	Leg

15E. Short Answer

1. a. Remove the shank
    b. Remove the butt tenderloin
    c. Remove the pelvic bone
    d. Remove the top round
    e. Remove the shank meat
    f. Remove the round bone and knuckle
    g. Remove the sirloin
    h. Remove the eye of round

2. a. Top round   d. Sirloin
   b. Eye round   e. Bottom round
   c. Knuckle   f. Butt tenderloin

3. ***Primal***   ***Subprimal/Fabricated Cut***   ***Menu Example***

   Ribs:

   a.   Hotel rack   Roast veal with Porcini
                                                    mushrooms

   b.   Rib chops   Braised veal chop with risotto

   c.   Rib eye   Broiled rib eye with Chipotle
                                                    sauce

   Any three of the following would be appropriate answers:

   Loin:

   a.   Veal loin   Roasted veal loin with wild
                                                    mushrooms

   b.   Loin chops   Sautéed veal chops with
                                                    mushroom sauce

   c.   Boneless strip loin   Roasted veal loin sauce
                                                    poulette

   d.   Veal tenderloin   Sautéed tenderloin with
                                                    garlic and herbs

4. Reference page 381.

15F. Matching I

1. f
2. a
3. b (d is appropriate for breast only)
4. e
5. c

15G. Matching II

a. 2
b. 1
c. 3
d. 2
e. 1
f. 3

15H. Chapter Review

1. True
2. True
3. False   (pp. 383, 388)   Sweetbreads are pressed to improve their texture.

4. False	(p. 386)	Émincé should be cut across the grain.
5. True		
6. False	(p. 383)	The thymus glands shrink in older animals.
7. True		
8. True		
9. False	(p. 380)	Restricting their movement prevents their muscles from toughening.
10. True		
11. True		
12. False	(p. 380)	Veal shoulder would be better ground or cubed and cooked using a combination cooking method.
13. True		
14. False	(p. 390)	It is braised to break down the tough meat found in the foreshank.
15. True		

15I. Putting It All Together

1. Veal are usually male because they are more expendable as livestock. A farmer wants to keep as many female cows through adulthood as possible because they are more valuable by extending the heard through giving birth to numerous offspring as well as producing milk. One male cow can serve as stud to hundreds of cows, thereby limiting the number that are needed on a farm at any one time.

2. Like beef, veal is a major source of protein as well as niacin, zinc and B vitamins. Veal has less marbling than beef, but because the animal is so young and its activity has been limited the meat is still tender. Due to the decreased marbling once visible fat is trimmed, veal is lower in fat and calories than comparable beef cuts. It is also leaner than many cuts of pork and poultry. Since cholesterol is found in the fat, with a lower fat content veal is also a bit lower in cholesterol.

3. As a younger and more tender animal the length of the cooking time involved in braising or stewing veal is considerably shorter than that for beef. With its mild flavor, braising and stewing will tenderize but will also make the flavors of the meat more complex than if it had been cooked using a dry cooking technique like sautéing or grilling.

➢ CHAPTER 16 ANSWER KEY

16A. Terminology

Answers will not be provided in answer key. All answers can be found in text.

16B. Primal Cuts Identification

Answers will not be provided in answer key. All answers can be found in text pages 406-409.

16C. Primal Cuts of Lamb

1. Shoulder
2. Rack
3. Loin
4. Leg
5. Breast

16D. Subprimal or Fabricated Cuts

Prismal Cut	Subprimal/Fabricated Cuts	Cooking Methods
1. Shoulder	a. Chops	Broil/grill
	b. Diced/ground	Stew/grill
2. Rack	a. Chops	Broil/grill
	b. Lamb rack	Roasted
3. Loin	a. Chops/boneless roast	Grill/roast
	b. Medallions/noisettes	Sauté
4. Leg	a. Lamb leg (bone-in)	Braise
	b. Boned leg	Roast
5. Breast	a. Breast	Braise
	b. Lamb shanks	Braise

16E. Cuts of Lamb and Applied Cooking Methods

Cooking Method	Subprimal/Fabricated Cut	Primal Cut
1. Broil/grill/roast	Loin chops	Loin
2. Stew	Diced lamb	Shoulder
3. Broil/grill/roast/sauté	Lamb loin	Loin
4. Stew	Diced lamb	Shoulder
5. Broil/grill/roast/sauté	Frenched lamb rack	Hotel rack
6. Combination (braise)	Breast	Breast
7. Dry heat (broil/grill/roast/sauté)	Lamb rack	Hotel rack
8. Dry heat (roast)	Lamb leg	Leg

16F. Short Answer

1. Reference pages 409-410.
2. Reference pages 410-411.
3. Reference page 411.

16G. Chapter Review

1. False   (p. 406)   Lamb primals are not classified into a forequarter and hindquarter as with beef, or a foresaddle and hindsaddle as with veal.

2. False   (p. 406)   Spring lamb is the term used to describe young lamb that has not been fed on grass or grains.

3. True

4. True

5. True

6. False   (p. 408)   The chine bone runs through the primal lamb rack.

7. True

8. True

9. False   (p. 406)   Frenching can be done on bones of other animals such as veal or pork.

10. False   (p. 425)   15 portions

11. False   (p. 409)   The rib eye comes from the rack.

12. True

13. True

14. True

16H. Putting It All Together

1. Even though Australian and New Zealand lamb accounts for 50% of the lamb sold in the U.S.A, the majority of lamb produced in our country is actually consumed here. Even though fewer Americans purchase lamb to prepare at home, sales of meals containing lamb in restaurants show that Americans enjoy eating it perhaps more than they like preparing it. In terms of the larger size of American lamb, it likely has to do with how they are raised. If American lambs are fed primarily grain and not grass like their competition it would suggest that the animal is less mobile during its short life, lacking the need to forage for its food. This, and likely the result of strong animal husbandry practices in America, lead us to fewer heads of animals produced, but better yield due to their larger size.

2. Lamb can not only be an economical source of red meat but it is also lean and lower in cholesterol than other red meat proteins and compared to chicken, fish or poultry it is a good source of iron. It is naturally lower in fat because it lacks the intensive marbling that other red meats possess, Since it is preferred young by the American public, before the animal matures and the flavor of its meat intensifies, it is naturally quite tender. And since most of the fat it gains settles on the outside of many cuts, it can be easily trimmed before cooking.

3. With knowledge and practice a chef should easily be able to break down the hotel rack him/herself, thereby saving a fairly significant amount of money per pound on a highly desirable cut of lamb on restaurant menus. A chef will know whether this savings is significant enough to continue the practice or if his/her time and money is too valuable. Can s/he break down the hotel rack for less money and if so, how much less?

➤ CHAPTER 17 ANSWER KEY

17A. Terminology

Answers will not be provided in answer key. All answers can be found in text.

17B. Primal Cuts of Pork

1. Boston Butt   4. Belly

2. Loin   5. Shoulder

3. Fresh ham

17C. Subprimal or Fabricated Cuts

Primal Cut	Subprimal Fabricated Cut	Cooking Methods	Cured & Smoked	Fresh
1.	Boston butt	Broil/grill/sauté	X	X
2.	Pork back ribs	Steam—grill		X
	Pork loin chops	Broil/grill		X
	Pork tenderloin	Sauté/roast/braise/broil/grill	X	
	Pork loin	Roast/braise		X
3.	Fresh ham	Roast/boil	X	X
4.	Spare ribs	Simmer—grill	X	X
	Bacon	Sauté/grill	X	
5.	Picnic shoulder	Bake	X	X

17D. Cuts of Pork and Applied Cooking Methods

Cooking Method	Subprimal/Fabricated Cut	Primal Cut
1. Dry heat (roast)	Fresh ham	Fresh ham
2. Moist heat (simmer)	Boston butt	Boston butt
3. Dry heat (broil/grill/sauté/roast)	Pork tenderloin	Loin
4. Combination (braise)	Pork loin	Loin
5. Combination (steam-grill)	Spare ribs	Belly
6. Dry heat (roast/bake)	Picnic shoulder	Shoulder
7. Dry heat (sauté)	Bacon	Belly

17E. Short Answer

1. Reference pages 435-436.
2. a. Shoulder
   b. Shoulder hock
   c. Boston butt—cottage ham
   d. Spare ribs—belly
   e. Pork belly—bacon
   f. Fresh ham

17F. Matching

1. d
2. f
3. a
4. b
5. c

17G. Chapter Review

1. False   (p. 433)   The Boston butt is located in the forequarter.
2. True
3. True
4. False   (p. 432)   The foreshank is called the shoulder hock.
5. False   (p. 434)   Center-cut pork chops are the choicest chops from the primal loin.
6. False   (p. 434)   Canadian bacon is made from the boneless pork loin.
7. True
8. False   (p. 432)   Hogs are bred to produce long loins.
9. False   (p. 432)   Picnic ham is made from the hog's primal shoulder.
10. True
11. False   (p. 432)   A variety of cooking methods are usually applied to pork.
12. False   (p. 434)   Spareribs are cut from the belly.
13. True

17H. Putting It All Together

1. While pork is a good source of protein and B vitamins, it is not as good a source of iron as lamb. In addition, pork is much more marbled with fat than lamb as well as possessing an outer layer on the cuts. In general, it is higher in fat, especially saturated fat, than lamb. Lamb is less marbled with fat and what fat exists is found more on the outside of the cuts, making it easier to trim away before cooking. It is also common for numerous pork cuts to be preserved by curing and/or smoking. While more of these products are available in low sodium varieties today, it provides an excessive amount of the nutrient sodium, requiring consumers to pay more attention to creating balance in their diet elsewhere and decreasing the total amount of preserved pork products they consume.

2. The ribs of pork contain a relatively small amount of meat (muscle) combined with an equal amount of fat. The nature of how the animal grows and moves around suggests that those muscles along the rib cage are working muscles, making the meat more tough. (Think about your own rib cage and how much the muscles surrounding it work to move your torso from one side to the other all day long.)   As a result, the moist cooking method of steaming or boiling is first used to break down the tough muscle fibers of the meat while simultaneously flavoring it with the cooking liquid. This process also helps to render out some of the fat so that the ribs are more palatable. Many people also finish ribs by grilling them at the end of the cooking process, usually with some acidic sauce or rub like barbecue, which further helps to tenderize and flavor the meat. The dry cooking method used at the end helps to caramelize the remaining sugars in the pork, adding to its palatability.

➤ CHAPTER 18 ANSWER KEY

18A. Terminology

Answers will not be provided in answer key. All answers can be found in text.

18B. Short Answer

1. Reference page 477.
2. Similarity: Overused muscles are more tough than underused ones.
   Difference: Red meat has marbling, poultry does not.
3. Reference pages 465-466.
4. Reference page 466.
5. Reference page 460.
6. Reference pages 483-494.
7. a. Be sure all work surfaces and equipment are clean.
   b. Avoid getting poultry juices in contact with other food.
   c. Anything coming in contact with raw poultry should be sanitized before it comes in contact with any other food.
   d. Cooked foods should never be placed in containers that were used to hold the raw product.
   e. Kitchen towels used to handle poultry should be sanitized before being reused to prevent cross-contamination.
8. Reference page 468.

18C. Matching

1. b
2. e
3. a
4. f
5. c

18D. Fill in the Blank

1. Protein, Myoglobin
2. 165° F, 170° F
3. Roaster duckling, dark, fat
4. White wine, lemon juice, herbs, spices, barbecue

18E. Multiple Choice

1. c
2. b
3. a
4. b

18F. Chapter Review

1. False	(p. 460)	Poultry fat has a lower melting point than other animal fats.
2. True		
3. True		
4. False	(p. 471)	Poultry that is left too long in an acidic marinade may take on undesirable flavors.
5. False	(p. 460)	The cooking time for dark meat is longer.
6. False	(p. 465)	Poultry should be frozen at 0° F/-18° C or below.
7. False	(p. 460)	The skin color of poultry is related to what it is fed.
8. True		
9. False	(p. 461)	Older male birds have less flavor than older female birds.
10. False	(p. 464)	Overcooking foie gras will cause it to melt away.
11. False	(p. 463)	A young pigeon is known as a squab.
12. False	(p. 464)	The gizzard is the bird's second stomach.
13. False	(p. 461)	A capon is a castrated male chicken.
14. False	(p. 460)	Poultry is placed into classes based on age and tenderness.
15. True		
16. False	(p. 463)	Ostrich is best cooked medium rare to medium.
17. True		
18. False	(p. 471)	Discard the marinade once it has been used.
19. False		Many variations of poultry and their preparations exist. See Chapter 2 for details on temperatures.
20. True		
21. True		
22. True		
23. True		
24. False	(p. 485)	The weight measure of clarified butter is not equal to the volume measure.
25. True		
26. True		
27. True		
28. False	(p. 463)	Ratite meat is usually cooked medium rare to medium.

18G. Putting It All Together

1. There are several things a chef needs to consider when planning a menu and costing recipes:

    Customer demand: Duck is all dark meat and contains large amounts of fat. As a result, a chef must know his/her customers' tastes very well in order to put this entrée on the menu. Prepared skillfully, duck can be a delicious addition providing it is something the guests desire.

    Yield: One four-pound duck will yield two adult servings while a four-pound roasting chicken will serve four people. The bone-to-meat ratio is much higher for duck than chicken and therefore a chef needs to plan accordingly when ordering ingredients for production and planning mise en place.

    Food cost: The cost per pound for duck is often more expensive than chicken. Combine this with the fact that the chef will only get two portions out of each bird compared to four from a chicken and the food cost for the base ingredient is already higher because s/he is paying more per pound for something that yields less meat.

    Ultimately the chef wants to make sure duck will be an entrée that customers will order on a regular basis so that s/he can ensure it will be worth the effort to prepare and stock the perishable ingredients required. If the dish will not make money for the establishment there is no sense putting it on the menu.

2. Since chicken is fairly mild in flavor, so should be the flavors used in the marinade. Once a chicken is broken down into its parts (breast, leg and wing), most marinating should take no more than a couple of hours to infuse adequate flavor. Choices of ingredients to use in the marinade are numerous although a few are provided below:

    Acid: lemon juice, white wine, rice wine vinegar, sherry

    Oil: canola oil, corn oil, olive oil

    Seasoning: basil, sage, garlic, scallion, thyme, soy sauce, mild mustard

➤ CHAPTER 19 ANSWER KEY

19A. Terminology

Answers will not be provided in answer key. All answers can be found in text.

19B. Short Answer

1. a. Sausages
    b. Forcemeats
    c. Pâtés
2. Reference page 541.
3. Reference page 537.
4. Reference page 541.

19C. Multiple Choice

1. b	5. d
2. c	6. c
3. b	7. a
4. a	

19D. Chapter Review

1. True
2. True
3. False   (p. 538)   Mature boar is one to two years old.
4. False   (p. 538)   Wild game can only be served by those who hunt and share their kill.
5. True
6. False   (p. 540)   Game is lower in fat and higher in protein and minerals than other meats.
7. False   (p. 537)   There is no marbling in venison flesh.
8. False   (p. 536)   Large animals are available only precut into subprimals or portions.
9. True
10. True
11. True
12. False   (p. 536)   The type, age and cut of the game animal will determine its tenderness and therefore the cooking technique that should be employed. Most game meat contains little marbling of fat and therefore must be cooked lightly (medium rare to medium) to maintain moisture.
13. True
14. False   (p. 481)   Traditionally wild game was usually marinated before preparation but today, with the availability of young, tender farm-raised animals, marinating is no longer a necessity.
15. True
16. True
17. True

19E. Putting It All Together

1. Like lean beef, and providing the cut of the bison is tender (such as the strip loin, sirloin, tenderloin), then any dry method should work well, such as sautéing, grilling or broiling. However, if the bison is lean and from an area of the animal that the muscle was well developed such as the shank, shoulder or round, then a moist or combination cooking method would be more appropriate.

2. The mid section or loin area.

➢ CHAPTER 20 ANSWER KEY

20A. Terminology

Answers will not be provided in answer key. All answers can be found in text.

20B. Multiple Choice

1.	b	8.	d
2.	d	9.	c
3.	c	10.	d
4.	a	11.	c
5.	d	12.	c
6.	b	13.	d
7.	a	14.	c

20C. Market Forms of Fish

1. drawn
2. pan-dressed/dressed
3. fillets
4. whole or round
5. steaks
6. butterflied
7. wheel or center cut

20D. Short Answer

1. a. They cook evenly
    b. They cook quickly
2. a. Translucent flesh becomes opaque
    b. Flesh becomes firm
    c. Flesh separates from the bones easily
    d. Flesh begins to flake
3. a. Shallow poach
    b. Sauté
    c. Broil
    d. Bake
4. *oily:*
    a. Trout
    b. Salmon
    lean:
    a. Bass
    b. Snapper

5. a. Scallops
    b. Lobster
    c. Shrimp
    d. Crab
6. a. Baked stuffed shrimp
    b. Oysters Rockefeller
    c. Baked stuffed lobster
7. a. They are naturally tender
    b. They cook relatively quickly
8. a. Eyes
    b. Gills
    c. Fins and scales
    d. Smell
    e. Texture
    f. Appearance
    g. Movement

20E. Chapter Review

1. False   (p. 573)   Only fish processed under Type 1 inspection services are eligible for grading.

2. False   (p. 573)   Fish and shellfish inspections are voluntary and are performed in a fee-for-service program.

3. True

4. True

5. False   (p. 572)   Maine lobsters have meat both in their tails and claws and are considered superior in flavor to all other lobsters. Spiny lobsters primarily have meat in the tail.

6. False   (p. 568)   Atlantic hard-shell clams are also known as quahogs.

7. True

8. False   (p. 594)   En papillote is actually an example of steaming.

9. False   (p. 573)   In general, shellfish have less cholesterol than lamb and other meats.

10. True

11. True

12. False   (p. 564)   The only market form that monkfish is sold in is the tail (fillet).

13. False   (p. 563)   Surimi is very low in fat and relatively high in protein. Because of processing techniques, however, it has more sodium and fewer vitamins and minerals than the real fish or shellfish it replaces.

14. True

15. True

16. False   (p. 596)   Poached items get their seasonings from the cooking medium, such as a court bouillon.

17. True

18. True

19. False   (p. 585)   It prevents sticking and helps leaner fish retain moisture.

20. True

20F. Putting It All Together

1. Referring to information provided in this chapter and your knowledge of principles of cookery from Chapter Ten, you should realize that the most commonly used cooking methods for seafood—broiling, grilling, poaching, sautéing and steaming—add little or no fat. Granted, a chef could choose to pan fry or deep fat fry fish and shellfish, but realize that there are numerous preparation options that help the chef to naturally keep this category of ingredients healthful, highly nutritious and delicious.

2. Based on knowledge gained from the chapters on professionalism, food safety and sanitation, menus and recipes and knife skills, a chef should be able to determine:

   a. The food service operations has the ability to utilize the bones and trim that cutting a whole fish produces

   b. The employees' ability to fabricate fillets, steaks or portions (including shucking shellfish). This includes considering the level of staffing and the time required for employees to complete this task.

   c. Storage facilities and adherence to HACCP standards

   d. The product's intended use

   Even though fish fabrication is relatively simple, these are all things that need to be considered.

➢ CHAPTER 21 ANSWER KEY

21A. Terminology

Answers will not be provided in answer key. All answers can be found in text.

21B. Multiple Choice

1.	d	7.	a
2.	b	8.	b
3.	b	9.	a
4.	d	10.	c
5.	b, c, d, f	11.	b
6.	c	12.	c

21C. Egg Identification

a. Shell

b. Yolk

c. White

d. Chalaza

21D. Coffees and Teas

1. a. Body
   b. Smell
   c. Acidity
   d. Flavor
2. Arabica

3. Robusta
4. 2 tbsp, 6 ounces
5. Coffee
6. Tea
7. Black, green, Oolong

21E. Chapter Review

1. True
2. False   (p. 658)   Shell color has no effect on the quality (grade), flavor or nutrition.
3. False   (p. 666)   When preparing French-style omelets the eggs are cooked without a filling, then tightly rolled onto a plate for service. The finished omelet can then be cut and filled as desired.
4. False   (p. 673)   The cooking surface should be 375° F.
5. False   (p. 661)   The egg whites should be brought to room temperature to maximize the volume when whipping.
6. False   (p. 659)   Eggs should be stored at temperatures below 40° F and at a relative humidity of 70-80 percent.
7. False   (p. 660)   Egg substitutes have a different flavor from real eggs and cannot be used in recipes where the eggs are required for thickening.
8. False   (p. 659)   Egg whites contain more than half of the protein and riboflavin, but no cholesterol.
9. False   (p. 673)   The pancakes should be golden brown and bubbles should be forming before they are flipped.
10. True
11. True
12. True
13. False   (p.663)   Custards are baked, not poached.
14. True
15. True
16. False   (p. 672)   Grids of waffle irons should be seasoned well and never washed. The heat of the iron keeps it sanitary.
17. True
18. False   (p. 659)   Egg whites coagulate at a lower temperature than yolks, so when preparing an egg white omelet, adjust your cooking time and temperature accordingly.
19. False   (p. 673)   Crepes are a type of pancake, however they are thin, delicate and unleavened.
20. True
21. False   (p. 682)   Green tea is yellow-green in color with a bitter flavor, but is not fermented at all.
22. False   (p. 680)   Cafe latte: mix 1/3 espresso and 2/3 steamed milk without foam.

23. True

24. True

25. False   (p. 676)   Bottled water comes from a variety of sources but whether from a spring, well or an artesian source, and whether it is still or sparkling, the bottle itself should be lightly chilled and served in a clean glass. Do not pour the water over ice unless requested.

26. False   (p. 681)   A variety of cultures have been flavoring their coffees long before the U.S.A. was even a country. For example, toasted barley, dried figs and spices are common flavoring ingredients and the French have been adding ground chicory root to their coffee for its added bitterness. In America we enjoy coffees that are flavored, often with aromatic oils, that add chocolate, vanilla, spice or raspberry flavors to name a few.

27. True

28. True

21F. Putting It All Together

1. Whole eggs, particularly the yolk where the fat and cholesterol are contained, do not have as much cholesterol as researchers once feared. Like everything else, eggs should be eaten in moderation and it is as much the saturated fat that is contained in the egg yolk that should be of concern. However, the American Heart Association suggests that it is acceptable to consume up to four egg yolks per week as part of a balanced diet. Of course, if one commonly eats a 4 to 5 egg omelet several times per week, then perhaps it is time to consider cutting back. One way of doing this is to use fewer egg yolks and more egg whites, which contain no fat or cholesterol and are a great, economical source of high quality protein. Don't forget to include in the computation of how many eggs eaten per week those that are "hidden ingredients" in baked goods, sauces, pastas and other common foods.

2. There are many things a chef can do to keep this potentially hazardous food product that is high in protein safe for human consumption:

   a. Keep all raw eggs consistently refrigerated until needed for preparation. Refrigerate at 41° F or below.

   b. Work cleanly and neatly in order to prevent cross-contamination.

   c. Keep eggs and egg mixtures well chilled over an ice bath when working with them at room temperature. For example, when cooking eggs to order, keep them on ice. Containers of cracked and whisked eggs for omelets and scrambled eggs should be brought out of refrigeration in small quantities and stored in a container that is fully submerged in an ice bath in order to keep it well chilled. Stir often to ensure that the center of the egg mixture is as cold as the outer edges. If few egg orders are being placed, return the eggs or egg mixture to formal refrigeration.

   d. Be careful to wash and sanitize cooking implements often. For example, do not use the same dirty spatula to make omelets during a 2 or 3 hour service. Switch out spatulas on a regular basis to ensure cleanliness.

   e. Cook eggs completely, so that yolks and whites are firm. Because eggs tend to carry salmonella, thorough cooking is one of the best ways of preventing the bacteria from being transferred to the guest.

   f. Serve hot, freshly cooked eggs and egg products immediately and at the proper temperature.

➢ CHAPTER 22 ANSWER KEY

22A. Terminology

Answers will not be provided in answer key. All answers can be found in text.

22B. Multiple Choice

1	b	7.	c
2.	c	8.	d
3.	b	9.	b
4.	c	10.	b
5.	a	11.	d
6.	c	12.	b

22C. Product Identification

1.	i	8.	d
2.	b	9.	m
3.	f	10.	l
4.	h	11.	c
5.	a	12.	e
6.	j	13.	n
7.	g	14.	o

22D. Chapter Review

1. True
2. False   (p. 742)   Pureed vegetables are usually prepared by baking, boiling, steaming or microwaving.
3. False   (p. 710)   Winter squash is generally not braised or stewed and is better cooked by baking, steaming or sautéeing.
4. True
5. True
6. True
7. True
8. True
9. False   (p. 725)   They are usually stored at 40-60° F.
10. True
11. False   (p. 728)   Red and white vegetables such as red cabbage, beets and cauliflower contain flavenoids.
12. False   (p. 729)   Testing the texture, looking for an al dente consistency, is generally the best determination of doneness.
13. False   (p. 725)   The ripening process of vegetables proceeds more rapidly in the presence of ethylene gas.

14. True

15. False   (p. 725)   The FDA classifies food irradiation as an additive.

16. False   (p. 704)   This technique is recommended when eggplant is going to be fried.

17. False   (p. 706)   A Chipotle pepper is a Jalapeño that has been smoked.

18. False   (p. 726)   Heat required in the canning process is what causes the contents of the can to lose nutrients and the texture to soften.

19. True

20. True

21. True

22. True

23. True

24. False   (p. 740)   Vegetables are generally tender; braising and stewing may be performed to obtain an exceptionally flavorful final product.

25. False   (p. 717)   When soaking dried beans, allow 3 cups of cold liquid for each cup of dried beans.

22E. Putting It All Together

1. Some simple rules to help you to remember how to get the most out of vegetables:

   a. Eat a variety of vegetables; the more colorful your plate, the better range of nutrients you're consuming.

   b. Choose fresh vegetables whenever possible and wash them thoroughly before preparation. If fresh are not available, then select frozen. If frozen are not available then choose canned vegetables as a last resort.

   c. Eat a variety to help ensure you're getting a variety of fiber (both soluble and insoluble) in your diet.

   d. Prepare vegetables with their peels intact whenever possible in order to preserve their nutritional content.

   e. Cook vegetables as lightly as possible if at all in order to maximize their nutritional value.

2. The information on color presented in Chapter Seven primarily talked about the appeal of vibrant and varied colors in the final presentation of food. As diners we form a certain opinion about how a food will taste simply based on how it looks. In many ways Chapter 22 talked about the same thing but in a more in-depth way. This chapter went on to explain how various vegetables and the pigments they contain react positively and negatively to reactions to acids and alkalis that may or may not be present during a cooking process. The affects on the vibrancy of the pigments was also related to the concept of what happens when a vegetable is overcooked.

➤ Chapter 23 Answer Key

23A. Terminology

Answers will not be provided in answer key. All answers can be found in text.

23B. Short Answer

1. To allow the pasta ample space to move freely and so that the starches that are released don't cause the pasta to become gummy and sticky.

2. a. Wrapping the potato in foil causes it to steam instead of bake and the skin will be soggy.
   b. Microwaving also causes steaming to occur and causes the skin to be soggy.

3. a. Italian risotto
   b. Spanish paella
   c. Japanese sushi

4. This gives the dough a rich, yellow color, and the dough is more resilient to the machinery during high-scale production. It also produces pasta that has a lightly pitted surface, causing the pasta to absorb sauces well.

5. a. Duchesse + tomato concassee = Marquis
   b. Duchesse + chopped truffles, almond coating and deep fried = Berny
   c. Duchesse + pâte à choux = Dauphine
   d. Dauphine + grated Parmesan, piped and deep fried = Lorette

6. a. Ribbon
   b. Tubes
   c. Shapes

7. a. Simmering
   b. Pilaf
   c. Risotto

8. a. The water softens the noodle strands.
   b. The bundles begin to separate.
   c. The noodles cook more evenly.

23C. Multiple Choice

1. b
2. c
3. d
4. d
5. c
6. b
7. b
8. b
9. b
10. a
11. a
12. a
13. d

23D. Chapter Review

1. True
2. True
3. True
4. False   (p. 793)   Semolina flour, although it makes the dough more yellow, also makes it tougher and more difficult to work with.
5. True
6. False   (p. 775)   A yam is botanically different from both sweet and common potatoes. Although it is less sweet than a sweet potato, it can be used interchangeably.
7. False   (p. 776)   Potatoes should be stored between 50 and 65° F.
8. False   (p. 776)   Waxy potatoes are best for these applications.
9. True
10. True
11. True
12. True
13. False   (p. 786)   The standard ratio for cooking rice is two parts liquid to one part rice.
14. True
15. False   (p. 786)   Generally cracked wheat and bulgur cannot be substituted for one another in recipes.
16. True
17. False   (p. 786)   What makes rice a potentially hazardous food product is its neutral pH and high protein.
18. False   (p. 788)   One cup of dried quinoa yields two cups cooked.
19. False   (p. 796)   Dumplings are defined as a small mound of dough cooked by steaming or simmering, usually in a flavorful liquid.
20. True
21. False   (pp. 777, 786)   Potato and grain dishes are indeed potentially hazardous foods but they should first be heated to 165° F and then held for service at 135° F.

23E. Putting It All Together

1. Potentially hazardous foods (PHFs) are termed as such because they contain protein, have a neutral pH and high water content. The reason there is such a heavy reminder of the dangers of PHFs in these two chapters is because it is easy to forget about this concept when talking about vegetables and carbohydrates. Many chefs have no problem relating the PHF definition to meats, eggs, seafood and dairy products but they often forget that vegetables and potatoes, grains and pastas can possess the same nutritional component of protein, making them equally susceptible to high rates of bacterial growth if not properly handled. Since bacteria feed on protein, need moisture and a neutral pH to thrive, all they need is temperatures within the temperature danger zone and the conditions are perfect for growth and reproduction thereby making vegetables, potatoes, grains and pastas equally susceptible to causing food-borne illness.

2. As explained in the earlier answer, potatoes are a potentially hazardous food product. Potatoes also grow in soil, soil that may be fortified with manure and other contaminants that may not be favorable to the human body when consumed. When a chef fails to wash and sanitize a potato before baking, then wraps it in aluminum foil, s/he is creating an environment where anaerobic bacteria can not only survive, but thrive. Botulism, for example, is not hindered by the application of heat and thrives in environments void of oxygen (while most bacteria require oxygen to survive). Wrapped inside the aluminum foil and containing a minimum of 60% water content, the potato provides plenty of oxygen and food for the bacteria to reproduce. The same thing can happen with a sealed container of rice pilaf or mashed potatoes that are being held in a holding oven to stay warm during service. Chefs need to be incredibly aware of everything they do in the kitchen in order to insure that food is safe for human consumption.

➢ CHAPTER 24 ANSWER KEY

24A. Terminology

Answers will not be provided in answer key. All answers can be found in text.

24B. Vegetarianism As Determined By Religion

1.	a, c	9.	d
2.	b	10.	c
3.	c	11.	c
4.	a	12.	b
5.	d	13.	b
6.	a	14.	a, c
7.	a, b, d	15.	a
8.	b		

24C. Short Answer

1. Vitamin: B-12
    Mineral: Iron
2. a. Religious beliefs
    b. Philosophical beliefs
    c. Environmental concerns
    d. Health concerns
3. *Diet for a Small Planet* explored vegetarianism from an environmental perspective, showing concern that the problem of world hunger is caused in great part by the wasteful use of agricultural resources on supporting animals raised for meat. She justified her claim by noting that in 1971 80% of the grain grown in the U.S. was fed to livestock.
4. a. Nuts
    b. Canola oil
    c. Flax and other seeds
    d. Soy products

5. a. Animal rights (meaning ethical treatment of animals)

    b. Strong proponent of vegetarianism

6. A chef cannot just assume that a guest who claims to be vegetarian only eats vegetables or grains since there are so many different variations on the diet. S/he really needs to ask specific questions directly of the guest to delineate the parameters for a meal s/he will prepare to meet the vegetarian guests' diet.

7. a. Greens

    b. Cruciferous vegetables

    c. Soy products

    d. Take supplements

8. White miso is mild and lightly sweet, containing a high percentage of rice. Red or dark miso contains a larger percentage of soybeans, is aged longer and has a stronger, saltier flavor.

24D. Chapter Review

1. True

2. False   (p. 842)   While vegetarianism is definitely growing in popularity in the U.S.A., approximately 6 million observe it as a lifestyle.

3. True

4. False   (p. 846)   Although tossed salads may be garnished with meat, tofu, legumes, cheese, nuts or seeds as part of a special order, they do not generally contain these ingredients. Vegetarians who are conscientious of making sure they get adequate protein in their diets may need to order these ingredients specifically.

5. False   (p. 848)   Eggs are only consumed by ovo or lacto-ovo vegetarians.

6. False   (p. 848)   Fruitarians do not consume grains.

7. True

8. False   (p. 848)   According to the Soy Bean Board, soy protein is the only plant protein that is equivalent to animal protein and it is a rich source of photochemicals as well.

9. False   (p. 848)   Once opened, the soy milk must remain refrigerated but only has a shelf life of 5-7 days.

10. True

11. False   (p. 850)   Cooking does temper tempeh's flavor, but it should be cooked due to the type of live culture it contains.

12. False   (p. 849)   While tofu can come in a variety of textures based on the amount of calcium sulfate added to the soymilk, it never quite takes on the texture of cooked textured soy protein or seitan, referred to by many as "wheat meat."

24E. Putting It All Together

1. Tofu may be bland, but because of this it is an extremely versatile, vegetarian ingredient that can be used in a plethora of dishes. It can be flavored in a variety of ways including marinating before cooking or incorporating it in a dish, such as a stir fry, where it picks up flavors from a variety of ingredients. As a result it should be seen as the ultimate protein-rich ingredient that can take on whatever flavor the chef desires.

2. First of all, one needs to consider the dish where substitution is taking place. For example, while making a veal stew, one can understand the texture the veal has upon completion of the stewing process (depending on the cut used in the recipe) and use that knowledge to choose the vegetarian protein ingredient that will most closely match the texture of the veal. The veal will be quite firm once cooked, so textured soy protein or seitan will probably most closely match the texture of veal. While the chef cannot find a substitute for the marbling of fat that exists in the grain of the veal that lends flavor and tenderness, at least s/he can utilize a protein that is similar to the firm texture. In addition, including other vegetable ingredients containing free glutamates like mushrooms, tomatoes and soy sauce, that add a savory or meat-like flavor can enhance the final characteristics of the adapted vegetarian stew.

➢ CHAPTER 25 ANSWER KEY

25A. Terminology

Answers will not be provided in answer key. All answers can be found in text.

25B. Multiple Choice

1. b	7. c
2. b	8. d
3. c	9. b
4. c	10. d
5. d	11. b
6. a	12. a

25C. Short Answer

1. a. Cheese and other high-fat dairy products
    b. Most meats (especially if high in fat)
    c. Most emulsified dressings
2. a. The gas causes the greens to wilt
    b. Accelerates spoilage
3. a. Buttermilk   d. Spices
    b. Vinegar   e. Vegetables
    c. Herbs
4. a. Bring mise en place up to room temperature.
    b. In the bowl of an electric mixer, whip the egg yolks until frothy.
    c. Add seasonings to the yolks and combine.
    d. Add a small amount of liquid from the recipe and combine.
    e. Begin whipping on high speed and slowly drizzle in oil until emulsion starts.
    f. After the emulsion forms, slow the mixer and add the oil a bit faster.
    g. When the mayonnaise is thick, add a small amount of the liquid from the recipe. Alternate this process with the oil until all incorporated.
    h. Taste, adjust seasonings and refrigerate immediately.

5. a. Liqueur

   b. Fruit puree

   c. Yogurt

   d. Sweetener, such as honey

25D. Chapter Review

1. True

2. True

3. False   (p. 874)   Tender greens such as butterhead and baby lettuces benefit from hand-tearing, while hardy greens like Romaine or Cos are acceptable to cut with a knife.

4. False   (p. 882)   Although tossed salads should in fact be dressed at the last possible moment, it is to prevent the greens from becoming soggy.

5. False   (p. 871)   Generally softer-leaved lettuces do tend to perish more quickly in storage than crisper-leaved varieties, however iceberg is not a soft-leaved lettuce.

6. True

7. True

8. False   (p. 877)   The standard ratio of oil to vinegar in a temporary emulsion is 3 parts to 1.

9. True

10. True

11. False   (p. 874)   Roses and zinnias are poisonous; chefs should be certain of the edibility of a bloom before using it to enhance food presentations.

12. True

13. False   (p. 885)   A conversion factor of 9.17 yields 55 portions.

14. False   (p. 879)   Use pasteurized egg yolks and keep ingredients and finished mayonnaise-based products below 41° F at all times.

15. True   Although the text only talks about a vinaigrette serving as a nice alternative for a light sauce, its ingredients are identical to a marinade as well (contains an acid, oil and seasonings). What a chef must avoid, however, is re-using a marinade as a sauce or dressing due to obvious food safety concerns.

25E. Putting It All Together

1. A chef has no formal obligation to offer vegetarian or low fat items, such as salads or salad dressings, for his guests' dining pleasure. However, from the standpoint of professionalism discussed in Chapter One and from the nutritional perspective in Chapter Three, the chef should get to know his/her clientele and do everything within reason to try to appease them. Success in today's food service industry is dependent on repeat customers, so if a chef can discover what his/her clientele desires or needs and strive to make them happy, then s/he as well as the establishment at which the chef is employed will benefit greatly. If we are ladies and gentlemen serving ladies and gentlemen, as the Ritz Carlton Hotel company credo suggests, then it is in our best interest to make the guest happy.

2. Salad Niçoise is dressed with a vinaigrette so most people would automatically assume that it is a healthful dining choice in and of itself. However, everything we eat must be taken in moderation and based on the nutritional analysis of the salad, perhaps the quantity of the ingredients (particularly the salad's garnishes) could be decreased in order to create a more healthful option. For example, 2 oz. of olive oil is rather heavy for a single serving of dressing, a whole tomato in addition to a variety of salad greens (about 4-6 ozs.), 4 ozs. of cucumber, 2 ozs. of green beans, a whole hard boiled egg (average weight 2 ozs.), a couple of artichokes, 2 ozs. of potato, green bell peppers, 4 ozs. of tuna (which is a very oily fish) and 1 oz. of olives could be considered an excessive amount of food, even for a dinner salad. If you add up the total ounces of a single portion a conservative estimate is 30 ounces—that's almost two pounds of food! This recipe might be good for an occasional splurge, but salads (and their portion size) need to be assessed before we can automatically rate them as a healthful dining option.

> ## CHAPTER 26 ANSWER KEY

26A. Terminology

Answers will not be provided in answer key. All answers can be found in text.

26B. Short Answer

1. a. Bananas
   b. Tomatoes
   c. Apples
   d. Melons
2. A grayish cast or color on the fruit.
3. *Vitamin C:* citrus, melons and strawberries
   Vitamin A: apricots, mangoes and kiwis
   Potassium: bananas, raisins, figs
4. Process the fruit into:
   a. Sauces
   b. Jams
   c. Jellies
   d. Preserves
5. a. Irradiation
   b. Canning
   c. Freezing
   d. Acidulation
   e. Drying
6. a. Apples
   b. Bananas
   c. Pears
   d. Peaches

7. a. Apples
    b. Cherries
    c. Pears
    d. Bananas
    e. Pineapples

26C. Fill in the Blank

1. Poaching
2. Apples
3. Gourd
4. Grapes
5. Batter

26D. Product Identification

1. b	6. g
2. e	7. a
3. d	8. k
4. h	9. f
5. I	10. j

26E. Chapter Review

1. True
2. True
3. False   (p. 940)   Sulfur dioxide is added to prevent browning and extend the shelf life.
4. False   (p. 939)   Freezing is generally one of the worst preserving methods for preserving the natural appearance since all fruits are 75-95% water which seeps out of the fruit when it defrosts.
5. False   (p. 937)   The highest grade is U.S. Fancy.
6. True
7. True
8. False   (p. 935)   Papayas are also referred to as Paw Paws.
9. True
10. True
11. False   (p. 929)   Red Delicious apples are best for eating raw.
12. False   (p. 931)   Although stone fruits are commonly dried or made into liqueurs and brandies, mangoes are not a stone fruit.
13. True
14. False   (p. 935)   Meat tenderizers contain enzymes similar to those found in pineapples and the seeds of kiwis and papayas.
15. True
16. True

17. False	(p. 942)	Fruits laid in a pan and sprinkled with a strudel topping and then baked are called crisps or crumbles.
18. False	(p. 918)	Berries must fully ripen on the vine as they will not ripen further after harvesting.
19. True		
20. False	(p.942)	Fruits with sturdy skins, such as apples and pears, work well for stuffing and baking.

26F. Putting It All Together

1. Citrus juices are highly acidic and play the role of the acid when making a marinade, tenderizing food but also adding flavor at the same time. Bacteria need a neutral pH in order to survive and thrive so when an acidic element is present, it will at the very least slow the bacterial growth down significantly. The oil also in the marinade will form a layer on top, preventing further incorporation of oxygen, which most bacteria also need to thrive.

2. Raw. Many nutrients, especially vitamin C, begin to diminish when heat is applied.

3. Organic production of food was introduced in Chapter One, Professionalism, presenting it as an option for chefs to consider when sourcing their ingredients and planning their menus. Organically produced produce is grown more naturally and while it might not be the prettiest, the taste is often superior. Heirloom produce, including fruits, offer chefs another option when they look to include the freshest, most original and highest quality ingredients on their menus. While the growing population of our planet benefits from our high volume farming methods that provide a consistent food supply, the quality and characteristics of produce can be lost, particularly when we consider how long the food sits in refrigeration before it is actually consumed. While not every chef will take advantage of the resurgence of heirloom varieties, it is a wonderful option for chefs to be able to explore the possibilities.

➤ CHAPTER 27   ANSWER KEY

27A. Terminology

Answers will not be provided in answer key. All answers can be found in text.

27B. Short Answer

1. Hot, cold
2. Protein
3. Human hands
4. Butter, mayonnaise, vegetable purées
5. Bound
6. Hot, open-faced
7. Reference page 966.

27C. Multiple Choice

1. a
2. c
3. d
4. a
5. a

27D. Matching

1. e
2. b
3. a
4. c
5. f

27E. Chapter Review

1. False (p. 961) Unlike butter, vegetable purées do not provide a moisture barrier between the bread and fillings.

2. True

3. False (p. 961) Butter, mayonnaise and vegetable purées are classified as spreads.

4. True

5. True

6. False (p. 967) For sit-down service hamburgers are often presented open faced for a more attractive presentation.

7. True

8. False (p. 967) A gyro is made with thinly sliced, rotisserie-cooked lamb wrapped in pita with onions and cucumber yogurt dressing.

9. False (p. 962) Keep cold foods cold (below 40° F) and hot foods hot (above 40° F).

10. False (p. 967) Thinly sliced corned beef is used in a Reuben, not roast beef.

27F. Putting It All Together

1. Sandwiches are extremely popular but like many other American meals can easily become nutritionally unbalanced. First, when looking at the nutritional guidelines provided by the Food Guide Pyramid (FGP), a traditional sandwich made with two pieces of bread actually represents two portions of bread, from the grains food group, not one. Therefore, if trying to eat according to the FGP one would either have to eat only half the sandwich or opt for an open-faced sandwich instead. A chef could also choose ingredients, particularly the sandwich meats, that contain a reduced level of fat and sodium. If consuming the entire sandwich then the portion size of the main component (meat, for example) should be reviewed to ensure that it does not exceed 4 ounces. Vegetables, which are often used as a garnish, should be used with abandon as they provide valuable vitamins, minerals and fiber.

Finally, the spread or dressing could be reviewed to determine whether or not it is a low fat, low sodium option.

2. Many of the ingredients in a sandwich are potentially hazardous food products due to the high percentage of protein they tend to contain. As a result the chef should follow HACCP standards closely, being sure to monitor time/temperature practices related to mise en place, assembly of the order and the proper packing of the foods "to go" utilizing ice packs that will maintain a temperature of the foods below 41° F.

> ## CHAPTER 28 ANSWER KEY

28A. Terminology

Answers will not be provided in answer key. All answers can be found in text.

28B. Short Answer

1. a. Basic forcemeat
   b. Country-style forcemeat
   c. Mousseline forcemeat
2. Add small quantities of crushed ice, bit by bit, to the machine while it is grinding.
3.

	Galantine	*Ballotine*
a.	Uses whole chickens, ducks, etc.	Uses poultry legs
b.	All bones are removed	All bones are removed
c.	Cavity of bird is filled with forcemeat	Cavity of leg is filled with forcemeat
d.	It is wrapped in skin, plastic, cheesecloth	Cooked without wrapping
e.	It is poached	It is poached or braised
f.	Always served cold	Usually served hot

4. a. Keep a precise ratio of fat to other ingredients
   b. Maintain temperatures below 40° F during preparation
   c. Mix ingredients properly
5. a. To glaze, preventing drying out and oxidation of food
   b. To cut into decorative garnish
   c. To add flavor and shine
   d. To bind mousses and salads
   e. To fill cooked pâtés on croute

28C. Multiple Choice

1.	a	6.	b
2.	c	7.	c
3.	a	8.	b
4.	a	9.	d
5.	d	10.	d

28D. Matching

1. f	6. a
2. i	7. g
3. e	8. c
4. d	9. j
5. b	

28E. Chapter Review

1. False   (p. 993)   Mousseline forcemeats can only be made out of meats, poultry, fish or shellfish.

2. False   (p. 996)   The best type of mold to use is a collapsible, hinged, thin metal pan.

3. True

4. False   (p. 987)   Eggs and egg whites are used as a primary binding agent in some styles of forcemeats.

5. True

6. False   (p. 987)   When marinating forcemeat ingredients before grinding, the trend today is to marinate them for shorter periods to let the natural flavors of the ingredients dominate.

7. True

8. True

9. False   (p. 1007)   A fresh ham is made from the pig's hind leg.

10. False   (p. 988)   After testing a forcemeat's texture and finding it too firm, a little cream should be added to fix the problem.

11. True

12. True

13. False   (p. 998)   Chopped chicken liver should be eaten within a day or two of its preparation whereas rillettes will keep for several weeks under refrigeration.

14. True

15. True

16. True

17. False   (p. 997)   Vegetable mousses are cold preparations made by combining puréed vegetables with béchamel, whipped cream and binding with aspic.

18. True

19. True

28F. Putting It All Together

1. Meatloaf and meatballs both contain similar ingredients. While we don't usually think of these common, everyday recipes as being categorized as forcemeats, they are—minus the fact that we don't emulsify the mixture. In general the concept of garde manger and forcemeats tends to be new, unfamiliar and even daunting to the young cook. But if one realizes that we

already understand the basics of forcemeats through the recipes we have made before, they may seem a little easier to comprehend. The simple technique of emulsification can create a completely different dish.

2. Common recipes for meatloaf and meatballs often contain a panada, but as laypersons we don't usually realize that we already understand this seemingly complicated principal—it's actually quite simple!

➢ CHAPTER 29 ANSWER KEY

29A. Terminology

Answers will not be provided in answer key. All answers can be found in text.

29B. Caviar

Reference pages 1030–1031.

29C. Multiple choice

1. d
2. b
3. c
4. b
5. d

29D. Short Answer

1. a. Fish
   b. Rice
   c. Seasonings
2. Reference page 1026.
3. Reference page 1027.
4. Reference page 1033.
5. Reference page 1029.
6. Reference page 1027.
7. Reference page 1030.

29E. Fill in the Blank

1. Brochettes
2. Pan-fried
   Deep-fried
3. Three, five
   Four, five
4. One

29F. Matching

1. d

2. c

3. a

29G. Chapter Review

1. False   (p. 1026)   Appetizers are usually the first course before the evening meal.

2. True

3. True

4. False   (p. 1031)   Caviar should be served in the original container or a nonmetal bowl on a bed of crushed ice.

5. False   (p. 1027)   Canapés with bread bases tend to become soggy so spreading butter on the base prevents this.

6. False   (p. 1027)   Canapés are best made as close to service as possible.

7. True

8. True

9. False   (p. 1033)   Rice wine and other seasonings are added to short-grain rice.

10. False   (p. 1035)   Filled pastry shells should be assembled at the last possible moment and then served immediately to prevent them from becoming soggy.

11. True

12. True

13. True

29H. Putting It All Together

1. Both stuffed, fried wontons and Rumaki are potentially hazardous foods due to their high protein content. And because they are being prepared in large quantities to be able to serve a large number of party guests over a considerable period of time, the chef not only needs to consider making sure the quality of the food is perfect, not under or overcooked, but s/he also needs to consider how far in advance s/he can cook the food and hold it for service. Considering the fact that Rumaki is made with chicken livers, which become very grainy and bitter when overcooked; and that fried wontons could easily become soggy if held to long, due to the accumulation of moisture that will make the crisp wonton wrapper become limp; the chef needs to consider that these hors d'oeuvres need to be cooked nearly to order in the kitchen rather than cooked as one big batch and held hot (250° F or higher) in a holding box or oven until needed for service. This will not only ensure that the food is safe to serve, but also enjoyable to consume.

2. Sushi is a particularly tricky hors d'oeuvre to serve, particularly on a buffet. Due to the protein found in the seafood, which is raw and ready to eat, and the protein in the sticky steamed rice, it is potentially hazardous because if not stored properly and consumed immediately it could easily become a breeding ground for bacteria, causing food-borne illness. The chef should set up a preparation station for the cook assigned to the station, either on the buffet line or in the back of the house where s/he can keep the seafood well below 41° F and the rice cool, but not cold. Preparation of the sushi in large quantities well before opening of the buffet and chilling it in the walk-in is not an option since while it keeps

the seafood cold and out of the temperature danger zone, the starch in the rice becomes very firm and gummy. Those who appreciate sushi properly prepared know that while the seafood should be kept ice cold and cut to order, the rice should be at room temperature in order to maintain a delicate and appropriate texture.

➤ CHAPTER 30 ANSWER KEY

30A. Terminology

Answers will not be provided in answer key. All answers can be found in text.

30B. Matching

1. d	6. e
2. I	7. a
3. f	8. j
4. b	9. g
5. c	

30C. Multiple Choice

1. d	6. a	11. d
2. b	7. d	12. b
3. c	8. d	13. a
4. b	9. c	14. c
5. d	10. a	15. a, d

30D. Short Answer

1. a. Melt in a container made from copper, aluminum or heatproof glass.

    b. Finely chop or grate chocolate to insure uniformity of melting.

    c. When using a double boiler, the water temperature should not exceed 140° F and the container holding the chocolate should not touch the water.

    d. Watch melting chocolate carefully and stir regularly.

    e. Remove the melting chocolate from the heat source when it reaches 115° F since carryover cooking will occur. Continue to whisk as the temperature of the chocolate rises to 120° F.

    f. Melt chocolate uncovered to prevent condensation build-up.

2. *Unsweetened*   *Bittersweet/Semisweet*

    a. Virtually inedible as is   35% chocolate liquor

    b. No sugar added   Sugar added

    c. No flavorings added   Flavorings added

    d. No emulsifiers added   Emulsifiers added

3. a. The process slowly raises and lowers the temperature of melted chocolate, preventing bloom.

   b. Causes chocolate to dry rapidly to a hard and shiny appearance.

   c. The chocolate shrinks as it dries, enabling it to be released from molds.

4. a. Spring form pans   d. Petite fours molds

   b. Tartlet pans   e. Various spatulas

   c. Piping tools

5. a. A lack of mixing and/or kneading of the dough

   b. If a formula contains too much fat in relation to the flour, the excessive fat has a tendency to coat the strands of gluten, preventing their development.

30E. Chapter Review

1. False   (p. 1068)   Self rising flour is all purpose flour with salt and baking powder added to it.

2. False   (p. 1066)   Glutenin and gliadin are the proteins, which when introduced to moisture and manipulated, form gluten.

3. True

4. False   (p. 1078)   Unsweetened chocolate is 100% chocolate liquor.

5. True

6. True

7. True

8. False   (p. 1067)   Flour derived from this portion of the endosperm is finer than other flours.

9. True

10. False   (p. 1068)   Unopened flour should be stored in the manner described, except it is also very important to store it away from strong odors, as it will absorb them easily.

11. True

12. True   In addition to these qualities, unsalted butter tends to be preferred because it is generally fresher than salted butter.

13. True

14. True

15. False   (p. 1073)   Most bakeshop ingredients combine completely with liquids, but fats do not.

16. False   (p. 1074)   Oils may not be substituted for solid shortenings in recipes.

17. True

18. False   (p. 1076)   The white coating is actually vanillin and the bean can still be used.

19. False   (p. 1079)   Do not substitute milk chocolate in any product that must be baked as the milk solids tend to burn.

20. False   (p. 1077)   The refining process for chocolate varies from country to country. For example, Swiss and German chocolate are the smoothest, followed by English chocolates. American chocolate has a noticeably more grainy texture.

21. True

22. True

30F. Putting It All Together

1. The first challenge is to determine what makes the bagel chewy. What element of flour makes it elastic? The elasticity is caused by the protein, also known as gluten, which is made more elastic the more the dough is kneaded. The second challenge is to go to your descriptions of the different flours available. Which ones contain the most gluten? Bagels are most commonly made from bread flour but occasionally, to increase the chewy characteristic of a bagel, chefs will often use a blend of bread flour and high gluten flour.

➢ CHAPTER 31 ANSWER KEY

31A. Terminology

Answers will not be provided in answer key. All answers can be found in text.

31B. Short Answer

1. The bitter or soapy flavor, and sometimes yellow coloring, is often caused by too much baking soda that may not have been properly mixed into the batter.

2. Baking soda can only release carbon dioxide to the extent that there is also an acid present in the formula. If the soda/acid reaction alone is insufficient to leaven the product, baking powder is needed for additional leavening.

3. Batters/doughs that may sit for some time before baking often use double-acting baking powder, which has a second leavening action that is activated only with the application of heat.

4. The higher fat content in the creaming method shortens the strands of gluten and therefore makes the final product more tender.

5. Softening the fat makes it easier to cream it with the sugar and therefore creates better aeration.

6. Overmixing the batter.

7. A scone is seen by many as a rich biscuit that also has butter and eggs in it. It is speculated that biscuits, at least the American form of the word, contain a less expensive type of fat, such as lard, and will omit the eggs.

8.   a.   Biscuits

      b.   Shortcakes

      c.   Scones

31C. Chapter Review

1. False (p. 1093) All-purpose flour is used in all of these methods.
2. True
3. True
4. False (p. 1090) Baking powder already contains both an acid and a base and therefore only moisture is needed to induce the release of gases.
5. False (p. 1090) All quick breads use chemical leavening agents and because they don't need to ferment, like yeast-leavened doughs, they are considered "quick."
6. True
7. False (p. 1092) Fats used in the muffin method should be in liquid form.
8. False (p. 1097) The leavening agent was there, so the assumption should be that the oven temperature was too low.
9. True
10. False (p. 1091) Batters and doughs made with single-acting baking powder should be baked as soon as they are assembled and mixed together.
11. False (p. 1090) Baking soda releases carbon dioxide gas if both an acid and moisture are present; heat is not necessary for leavening to occur.
12. False (p. 1091) Shortcakes are made using the biscuit method.
13. True
14. False (p. 1093) Muffin and quick bread batters are similar and therefore their baking methods are interchangeable as long as the baking time is altered.
15. True

31D. Putting It All Together

1. The basis of the question is really based on common sense but also reminds us of the importance of weighing certain ingredients while we do a volume measure of others. In this scenario blueberries weigh more due to their water weight while pecans are lighter.
2. The two ingredients that occur in the greatest quantity are the flour and the butter; nearly a 1:1 ratio—that's a lot of butter! Butter breaks up the strands of gluten in the flour, giving it flavor and tenderness that when caramelized with the sugars that are baked may have a faint firmness to it when it browns in the oven but crumbles in our mouth as we bite into it.

➢ CHAPTER 32 ANSWER KEY

32A. Terminology

Answers will not be provided in answer key. All answers can be found in text.

32B. Multiple Choice

1.	c	6.	a
2.	d	7.	d
3.	c	8.	a
4.	b	9.	c
5.	b	10.	a

32C. Short Answer

1. a. The yeast, liquid and approximately one half of the flour are combined to make a thick batter known as a sponge, which is allowed to rise until bubbly and doubled in size.

    b. Then the salt, fat, sugar and remaining flour are added. The dough is then kneaded and allowed to rise again. This creates a different flavor and a lighter texture than breads made with the straight dough method.

2. The organism is considered dormant because virtually all of the moisture has been removed, which helps to increase the shelf life, among other things.

3. a. Product size

    b. The thermostat's accuracy

    c. Crust color

    d. Tapping loaf on the bottom and listening for hollow sound

4. a. Croissants

    b. Danish pastries

    c. Non-yeast-leavened pastry

5. Halve the specified weight of compressed yeast when substituting dry yeast in a formula.

6. Combine all ingredients and mix.

7. a. Scale ingredients   f. Round portions

    b. Mix and knead dough   g. Makeup: Shape portions

    c. Ferment dough   h. Proof products

    d. Punch down dough   i. Bake products

    e. Portion dough   j. Cool and store finished products

8. a. 190° -210° F.   b. 180°-190° F

32D. Chapter Review

1. True   More specifically though, it occurs just after fermentation.

2. False   (p. 1113)   Salt's primary role in bread making is conditioning gluten, making it stronger and more elastic.

3. True

4. True

5. False   (p. 1118)   Underproofing results in poor volume and texture.

6. True

7. True

8. False   (p. 1113)   Active dry yeast contains virtually no moisture.

9. True

10. False   (p. 1113)   Prior to commercial yeast production bakers relied on starters to leaven their breads. Today starters are generally used to provide consistency and reliability.

11. True

12. False   (p. 1116)   Overkneading is rarely a problem.

13. False   (p. 1112)   Yeast is in fact very sensitive to temperature but it prefers temperatures between 75° F and 95° F.

14. True

15. True

16. False   (p. 1113)   All yeasts are destroyed at 138° F.

17. False   (p. 1117)   The fermentation process starts when the dough is finished mixing and continues until the dough is baked and reaches a temperature of 138° F and the yeast dies. This explains why the kneading, fermenting and punching down steps in the process are so important. Providing the formula is accurate and the ingredients used are correct, the gluten in the bread dough must be developed to its maximum ability to capture the gases produced from fermentation because no more rising will take place during the baking process (for the most part). Once bread goes into the oven and reaches an internal temperature of 138° F, the yeast dies and the ingredients of the dough are baked into the shape/structure they currently hold. Therefore, a dough that is under or over-proofed will not improve in quality when it goes into the oven—the quality is "frozen in time"—the resulting finished product will possess the same quality, usually rather dense, with the addition of the baking and resulting browning that occurs during the cooking process.

32E. Putting It All Together

1. Most breads that the American public loves to consume are in fact made from flour that is milled from the grain, wheat. Even though all-purpose flour, cake flour, pastry flour, bread flour and high gluten flour don't have the word "wheat" in their name like whole wheat flour does, they are all still derived from the same grain and therefore no, a person allergic to wheat could not eat bread as most Americans know it.

    There are alternative flours that can be used to make bread for those allergic, such as soy, rice, potato, tapioca, sorghum and cornstarch. Unfortunately, these flours contain no gluten to capture the gases produced during the fermentation process of yeast, and therefore an ingredient called Xanthan gum must be added to the formula. Even with this addition the resulting bread is considerably more dense than traditional wheat breads and many persons with this allergy have a difficult time enjoying breads in the same way that they once did.

2. In the process of assembling and mixing ingredients for a formula of bread dough the yeast is brought to the perfect temperature (usually by mixing with a liquid of the proper temperature—75°-95° F) to ensure activity, given food to eat (sugar) and then mixed with the remaining ingredients, which include flour. As fermentation progresses gases are produced. Mixing and kneading the dough develops the strands of gluten (making the protein increasingly elastic), which enables the dough to capture the gases produced, allowing the bread to rise (almost the same way a balloon captures our breath as we blow into it). The bread is then rounded and shaped, proofed to allow the fermentation process to be completed, and then baked to capture the dough in its light, airy form and to make it edible.

➢ CHAPTER 33 ANSWER KEY

33A. Terminology

Answers will not be provided in answer key. All answers can be found in text.

33B. Short Answer

1. a. Chiffon
   b. Cooked juice
   c. Cream
2. a. Cream
   b. Chiffon
   c. Cooked juice
   d. Cheesecake
3. a. Baked fruit
   b. Custard
4. a. Lattice coverings
   b. Pie top crusts
   c. Prebaked shells later to be filled with cooked fillings
5. It is a rich, non-flaky and sturdier dough than flaky or mealy dough due to the addition of egg yolks and the blending of the fat.

6. When the crust has a potential of becoming soggy, as in the making of custard and cooked fruit pies.

7. One can have better control because you can feel the fat being incorporated and therefore prevent overmixing.

8. It is cooked before baking.

9. The ratio of sugar to egg whites.

10. a. Vol au vents   c. Feuilletées

     b. Napoleons   d. Bouchées

11. a. Ratio of ingredients in dough

     b. Oven temperature

     c. Pan coating

33C. Multiple Choice

1. d   4. c

2. b   5. d

3. b

33D. Chapter Review

1. True

2. False   (p. 1160)   Strawberries, pineapples and blueberries would be more appropriate.

3. True

4. True

5. False   (p. 1157)   A typical ratio for crumb crusts consists of one part melted butter, to two parts sugar, to four parts crumbs.

6. True

7. False   (p. 1153)   Pâte sucrée should be used specifically over flaky and mealy doughs because it is less flaky and due to the addition of the egg yolks, is still tender, but is stronger to withstand the removal of the tart pan during service.

8. True

9. False   (p. 1160)   A cooked juice filling should be combined with a prebaked or crumb crust.

10. False   (p. 1163)   Baked fruit pies may be stored at room temperature until service.

11. True

12. False   (p. 1155)   An American gâteau refers to any cake-type dessert.

33E. Putting It All Together

1. The pie dough may take on a rancid flavor, particularly if it is made with a non-hydrogenated fat. In addition, a chef must take into consideration weather conditions where s/he works. For example, a geographic location that regularly maintains a relatively high humidity level in the air (perhaps 20% or greater) will cause the baked pie dough to also become soggy, losing its distinctive crunch and texture.

2. Cream and custard fillings, even once baked, are potentially hazardous foods. Therefore they must be stored at a temperature below 41° F until service. Combined with the potential moisture in the refrigerator, the condensation that is created on the pie/pie crust when bringing it from a cold temperature to a warm temperature and back again, and the inherent moist quality of the filling, the crust of a custard or cream filled pie is more likely to become soggy as a result of all these factors. As a result, the shelf life to maintain quality of the overall pie (provided it is refrigerated as directed) is realistically 1-2 days even though it is still safe to eat after 4-6 days.

➢ CHAPTER 34 ANSWER KEY

34A. Terminology

Answers will not be provided in answer key. All answers can be found in text.

34B. Basic Cake Mixes

Butter cake reference page 1214.
Genoise cake reference page 1216.
Sponge cake reference page 1218.
Angel food cake reference page 1219.

34C. Matching I—Ingredient Functions

1. f	4. c
2. a	5. d
3. e	6. g

34D. Matching II—Frostings

1. d	4. e
2. f	5. a
3. b	6. g

34E. Cake Mixing Categories

1. b	6. a
2. b	7. a
3. a	8. b
4. b	9. a
5. a	10. a

34F. Short Answer—Frostings

1. Simple buttercream reference page 1227.
2. Italian buttercream reference page 1228.
3. French buttercream reference page 1229.

34G. Fill in the Blank

1. Decreased, under whipped, increased, 25

2. a. Appearance

   b. Touch

   c. Cake tester should come out clean

3. 325, 375

4. Decorators' icing

5. Flour, shortening, oil

6. Combine the ingredients uniformly, incorporate air cells and develop the proper texture.

34H. Chapter Review

1. False   (p. 1223)   All cakes should be cooled away from drafts or air currents that could cause them to collapse.

2. True

3. True

4. True

5. False   (p. 1219)   Angel food cake is usually not frosted but may be topped with fruit-flavored or chocolate glazes, fresh fruit, fruit compote or whipped cream.

6. True

7. False   (p. 1242)   The results from package mixes are consistent and acceptable to most customers.

8. False   (p. 1215)   High-ratio cakes require emulsified shortenings to absorb the large amounts of sugar and liquid in the formula.

9. True

10. False   (p. 1226)   A compote is not a frosting although it may top or accompany a cake.

34I. Putting It All Together

1. Regarding the mixing and baking process there are not too many detailed food safety issues the chef needs to be concerned with when dealing with a butter, angel, sponge, chiffon or other type cake. As long as the ingredients are properly stored and handled and no contamination occurs from chemicals or hazardous objects, the process is pretty straightforward. If dealing with a cake that contains many potentially hazardous food ingredients, such as cheesecake, then much closer controls must be followed, including storing the cake at 41° F or below and protecting from cross-contamination until served.

   In the case of the butter, angel food, sponge, chiffon and other similar cakes, the area the chef needs to be most aware is once the cake is baked. Since it is a ready-to-eat product from this point on until service, care must be taken to prevent cross-contamination during storage, slicing and decorating stages. The same principles apply to the icing. As long as the icing or frosting contains no potentially hazardous food ingredients such as milk, cream, eggs or cream cheese, then it may be stored, covered, for several days at room temperature without cause for concern.

2. The food service establishment needs to determine whether or not they have the time needed for production, the talent required to produce a variety of desserts, the work and storage space, as well as the customer demand for house-baked products. If these things are answered and it seems economically advantageous, making desserts in-house could help the restaurant or other establishments create a niche.

➢ CHAPTER 35 ANSWER KEY

35A. Terminology

Answers will not be provided in answer key. All answers can be found in text.

35B. Short Answer

1. Sanitary guidelines for eggs reference page 1259.
2. Vanilla custard sauce reference page 1259.
3. Ice cream reference page 1273.
4. Baked soufflés reference page 1266.
5. Sabayon reference page 1262.
6. Mousse reference pages 1270–1.
7. Dessert assembly reference page 1278.
8. Precautions for ice cream reference page 1273.

35C. Fill in the Blank

1. Zabaglione
2. Mousseline, Italian meringue
3. Bavarians, Chiffons, mousses, crèmes chantilly
4. Egg whites will whip to better volume, two mixtures are more easily incorporated

35D. Chapter Review

1. False   (p. 1276)   A coulis sauce may be cooked or uncooked.
2. True
3. True
4. True
5. True
6. True
7. False   (p. 1274)   The sorbet may be soft and syrupy due to too much sugar in the formula.
8. False   (p. 1259)   There are several steps that can be followed to repair a curdled vanilla custard sauce.
9. False   (p. 1274)   A sherbet contains milk and/or egg yolks for creaminess.
10. True

35E. Putting It All Together

1. The process of adding the whipped egg whites to the other ingredients, particularly the chocolate base, involves using a long thin straight edge, such as that on a large rubber spatula, to incorporate the light, airy egg whites into the heavier base ingredients. This is done by scooping the whipped egg whites on top of the base in the bowl, then using the thin edge of the spatula, cutting through the two from one side of the bowl to the other, scooping up from the bottom to the top and repeating as the bowl is turned and until the two mixtures are blended lightly. By doing this with the thin edge of the spatula as opposed to the wide side of the spatula, and with a very specific movement, the chef is less apt to beat the air out of the egg white cells, therefore keeping them in place to use as leaveners during the cooking process.

2. A mousse generally has a light and airy texture yet a refined flavor, such as chocolate. If you've never had mousse before, in some ways it is an elegant version of pudding. The reason the text gives the chef the option of whether or not to add gelatin is based on the chef knowing how the mousse will be served to the guest. If gelatin is added, it is still used in small proportions in relation to the other ingredients because the mousse should still have, even if it contains gelatin, a smooth, light and airy texture (caused by the inclusion of whipped cream in the recipe) not rubbery. The gelatin simply aids in keeping the air molecules in the mousse until the dessert is consumed. So as an example, if a chef is serving the chocolate mousse in a martini glass topped with a mixture of seasonal berries, generally speaking, gelatin would not be required as the glass itself would serve to hold the mousse in place until it is consumed. However, if the mousse needs to be piped or is used as a filling between thinly sliced layers of a chocolate genoise, then the chef may choose to add a very small amount of gelatin to the mousse as s/he is making it so that the mousse is slightly more firm, helping to maintain the structure of the assembled cake once it is frosted and decorated.

3. Like similar questions that you've found in this study guide, the answer really comes down to balancing the time, talent and resources available to produce the product in-house. What makes this question different is that there really are a wide variety of quality ice creams, sherbets and sorbets available from retailers which really requires the chef to determine whether or not it is worth it for a food service establishment to produce its own. If ice cream is a signature item that guests drive from miles around to the restaurant to enjoy as a primary dessert or if the establishment creates custom flavors that accompany signature desserts that the restaurant is widely known for, then perhaps it is worth the investment. On the contrary, if ice cream is only used as a small garnish on a handful of desserts that sometimes sell but not consistently, then the decision is much simpler. A large ice cream machine, one that has a self-contained refrigeration unit, takes up a significant amount of space in a kitchen and can also be fairly expensive to purchase considering it can only produce a fairly limited array of desserts, is an unjustified expense in the latter scenario.

➤ CHAPTER 36 ANSWER KEY

36A. Terminology

Answers will not be provided in answer key. All answers can be found in text.

36B. Fill in the Blank

1. Hippen masse
2. Texture, shape, color
3. Size
4. Focal point
5. Cold

36C.   Short Answer

1. Cutting, molding
2. a. Flavor   d. Color
   b. Moisture   e. Texture
   c. Flow
3. a. Height
   b. Texture
4. a. To show the chefs' attention to detail
   b. To provide visual appeal
   c. To ensure even cooking of the product
5. a. Strike a balance between overcrowding and leaving empty gaps on the plate
   b. Choose a focal point
6. Make sure the plate's composition flows naturally
   a. Create height
   b. Add a new shape
   c. Keep plate neat and clean

36D. Chapter Review

1. True
2. False   (p. 1298)   The food should always be the focal point of any plate.
3. True
4. False   (p. 1304)   Dusting of a plate should be done before the food is plated.
5. False   (p. 1307)   A squeeze bottle would be a good choice of equipment for preparing sauce drawings.
6. False   (p. 1306)   An equally important concept is that the sauces need to be thick enough to hold a pattern and all sauces used in the drawing need to be the same viscosity.
7. True

8. False   (p. 1304)   Plate dusting is most commonly associated with pastry presentations.

9. True

10. False   (p. 1304)   Equally important is to decorate the presentation quickly so the food on the plate is served at its proper temperature.

36E. Putting It All Together

1. The chef will need to ensure that the recipe for the food is perfect and tested prior to service so that when it comes time to shape the food for the plate presentation there is no delay. S/he also needs to make sure that the cook who will be shaping, cooking and plating the quenelles to order is proficient at the skills involved. Stovetop space must be adequate as to ensure one burner can be dedicated to keep the poaching liquid (court bouillon) at the proper temperature during the entire service. Finally, the service team must be proficient at cooking foods to order and quickly and efficiently plating it so that hot food is served hot and cold foods are served cold.

2. There are several things the chef must consider once s/he has determined that the kitchen equipment line and space is conducive to a la minute plating and that the kitchen staff is willing and capable to complete the plating efficiently and effectively for each order. Common rules to focus on might include:

    a. Cost of garnish must be considered/limited, based on both the time it takes to make and perform as well as the cost of the ingredients. Garnishes not only need to complement the plate of food's flavors and enhance the presentation, but they must also be kept within budgetary constraints.

    b. The plating process must be efficient so that hot foods are served piping hot and cold foods are served ice cold. If the proper focus is not kept it is easy to let priorities slip and take too long to plate the food. In the process of excessive handling we also run the risk of cross-contaminating the finished food product before it is consumed by the guest.

➢ CHAPTER 37 ANSWER KEY

37A. Terminology

Answers will not be provided in answer key. All answers can be found in text.

37B. Short Answer

1. a. Offer dishes featuring different principal ingredients.
    b. Offer foods cooked by different cooking methods.
    c. Offer foods with different colors.
    d. Offer foods with different textures.
2. Buffet presentation reference pages 1145-1148.
3. a. Use a double-sided buffet.
    b. Use a single-sided buffet, divided into two, three or more zones, each of which offers the identical foods.
    c. Divide the menu among various stations, scattered throughout the room, each station devoted to a different type of food.

4. a. Choose foods that hold well.

    b. Cook small amounts of delicate foods.

    c. Ladle a small amount of sauce in the bottom of the pan before adding sliced meats.

    d. Keep chafing dish closed whenever possible.

37C. Multiple Choice

1. c

2. b

3. d

4. c

5. c

37D. Chapter Review

1. True

2. False   (p. 1312)   Costs are a consideration but the principal factors limiting a menu are the client's desires and the chef's imagination.

3. False   (p. 1314)   The buffet should be in an area with easy access to both the kitchen and the dining tables.

4. True

5. True

6. False   (p. 1323)   Portioning of these foods with potentially dripping sauces will be easiest for guests if the product is placed at the front of the table.

7. False   (p. 1322)   Try to avoid too much dead space on a buffet by filling in with decorations and props.

8. True

9. False   (p. 1325)   Spaghetti will not hold particularly well on a buffet but it also may be messy to serve—especially if topped with a sauce.

10. True

11. True

12. True

37E.   Putting It All Together

1. The Italian banquet that developed in concept during the Renaissance Period (1450-1600) was the first of its kind and therefore foods that were prepared for it were quite experimental compared to today. The practice of cooking foods to make them more palatable and safe to eat was in its infancy, so you can imagine that the fanciful preparation of foods was experimental at best. Since the concept of sanitation had not fully been defined at the time, food safety practices were pretty much nonexistent. It was not uncommon for banquets to last for days on end, and since refrigeration and the mechanical production of ice did not occur until the 1900's, you can imagine that the time/temperature concept was grossly abused—both in the preparation of the foods as well as in the service of them on the buffet.

Take a step forward in time to the modern buffet, in America, France, Belgium or any handful of developed countries, and you'll find quite a different picture, as is outlined in the text. Our cooking profession has come a long way in a relatively short period of time, not the least of which is the study of sanitation practices and how we as chefs can ensure that food is safe to consume.

2. Assign kitchen or dining room staff, professionally dressed in pristinely clean and pressed uniforms, to serve the food to guests. Such a practice:

   a. Ensures prompt service and a steady flow of traffic on the buffet

   b. Helps to prevent guests serving themselves and therefore better control portion sizes and prevent cross-contamination from utilizing one service utensil for two or more food items

   c. Helps to ensure the buffet table is carefully monitored; the presentation of the whole table stays neater and fresher with staff readily available and guests are less likely to play with the display or get unusually close to it, perhaps sneezing or coughing on it.

NOTES

NOTES

NOTES

NOTES

NOTES

NOTES

NOTES